# Contents

| | |
|---|---:|
| Foreword | 3 |
| Introduction | 4 |
| Fast ships at all times | 7 |
| Speed at any price | 13 |
| Fast combat boats after the war | 21 |
| Form theory and driving forces | 29 |
| When fast boats learn to fly | 35 |
| Small warriors | 40 |
| Riding on air | 46 |
| Highspeed in military logistics | 53 |
| Ships as from another star | 65 |
| America's invisible warriors | 75 |
| The view into the future | 86 |
| The far future of military ships | 90 |
| High speed boats in museums | 92 |
| Tips for model makers | 95 |
| Epilogue | 96 |

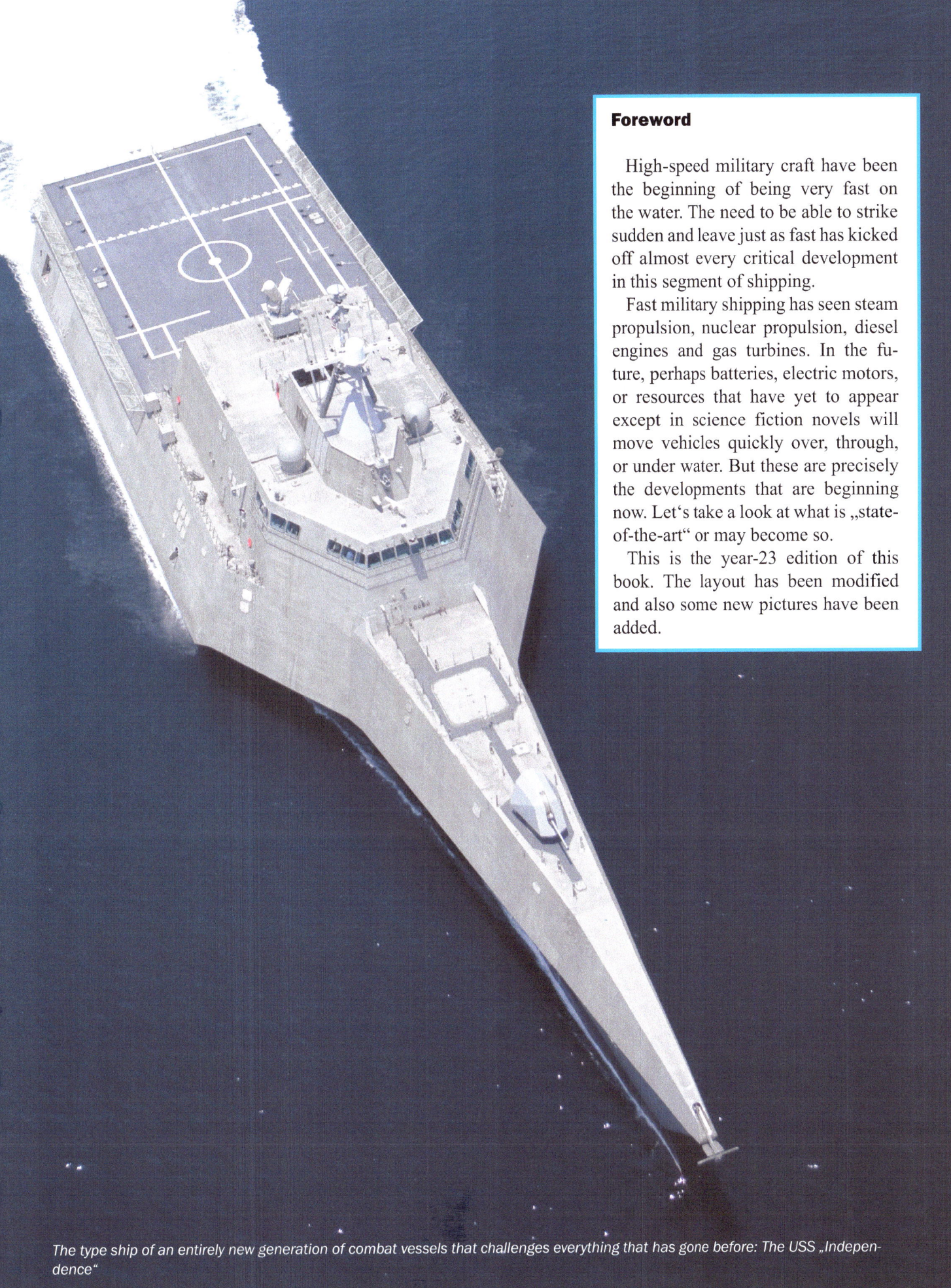

## Foreword

High-speed military craft have been the beginning of being very fast on the water. The need to be able to strike sudden and leave just as fast has kicked off almost every critical development in this segment of shipping.

Fast military shipping has seen steam propulsion, nuclear propulsion, diesel engines and gas turbines. In the future, perhaps batteries, electric motors, or resources that have yet to appear except in science fiction novels will move vehicles quickly over, through, or under water. But these are precisely the developments that are beginning now. Let's take a look at what is „state-of-the-art" or may become so.

This is the year-23 edition of this book. The layout has been modified and also some new pictures have been added.

*The type ship of an entirely new generation of combat vessels that challenges everything that has gone before: The USS „Independence"*

# Introduction

*"To be faster than anything that is stronger, or to be stronger than anything that is faster"* has been a common motto of particularly fast warships. First it was torpedo boats, then the first generations of destroyers, and later this phrase also applied to the missile speedboats of our time.

Today, they are super-fast watercraft, usually straight out of a science fiction novel, with ‚stealth' characteristics, state-of-the-art equipment, weapons and networked computer systems. Until the end of World War II, only a few battleships could reach speeds of 30 knots.

Speed was an important survival factor, especially for small and weakly armoured units. This was because in the days of artillery aimed by „eye," guns could not follow fast attackers as easily as today's high-powered guns guided by computers.

Today, speed is more of a „must" in order to reach a mission location quickly. The fast attack crafts seem to be obsolete in today's world. They are gradually being replaced in the German Federal Navy and other navies by the larger but slower corvettes.

Yet the „speedboat" genus has a history unlike any other naval warfare asset. They have taken the fight to the enemy, while the vast majority of other battleship genres have had to take a more defensive role. This has not changed to this day, as almost all major warships are tasked with protecting something, whether from air attack or attack by submarines.

The aggressive nature of a fast patrol boat stems from its role of really only being a carrier for offensive weapons such as missiles or torpedoes. It is not armoured and its defensive systems are too weak to effectively defend a single boat. It can only save itself from annihilation by its speed and raiding tactics.

In the 21st century, the Western world faces a new threat. In this world of worldwide use of warhips as police and peace keepers, the fast boats seem to have no place. Far from it, they have now mutated into a completely new type of vessel that can operate very effectively even when unarmed.

To do this, the fast craft had to become bigger and faster and completely change its shape. The future will bring new forms of propulsion, new fuels, drones above and below the water, and also artificial intelligence - and faster than most expect.

Technical data will only be mentioned in exceptional cases in this book. The Internet offers a very fast and always up-to-date source of information for this - better than a book ever could.

*A sign of times is this spectectular trial shot of a railgun in a US Navy test facility. The bullet is accelerated to a speed between Mach 6 and 7.5 - much faster than a 30-round of a rifle. The introduction of the railgun on US Navy ships will be revolutionary for the naval warfare.*

A fighter from the Cold War era: this former Israeli Saar IV FAC is now in service with the Chilean Navy. The Israelian 'Saar III' class are very capable boats and where very welcome in Chile

During World War II, this German E-Boat flotilla leaves for another mission in the English Channel. The fights between the Britsh and the German coastal forces where extremely hard. The main objective of the Germans was to interupt the transport of goods and fuel along the south coast of England.

PT 723 as completed

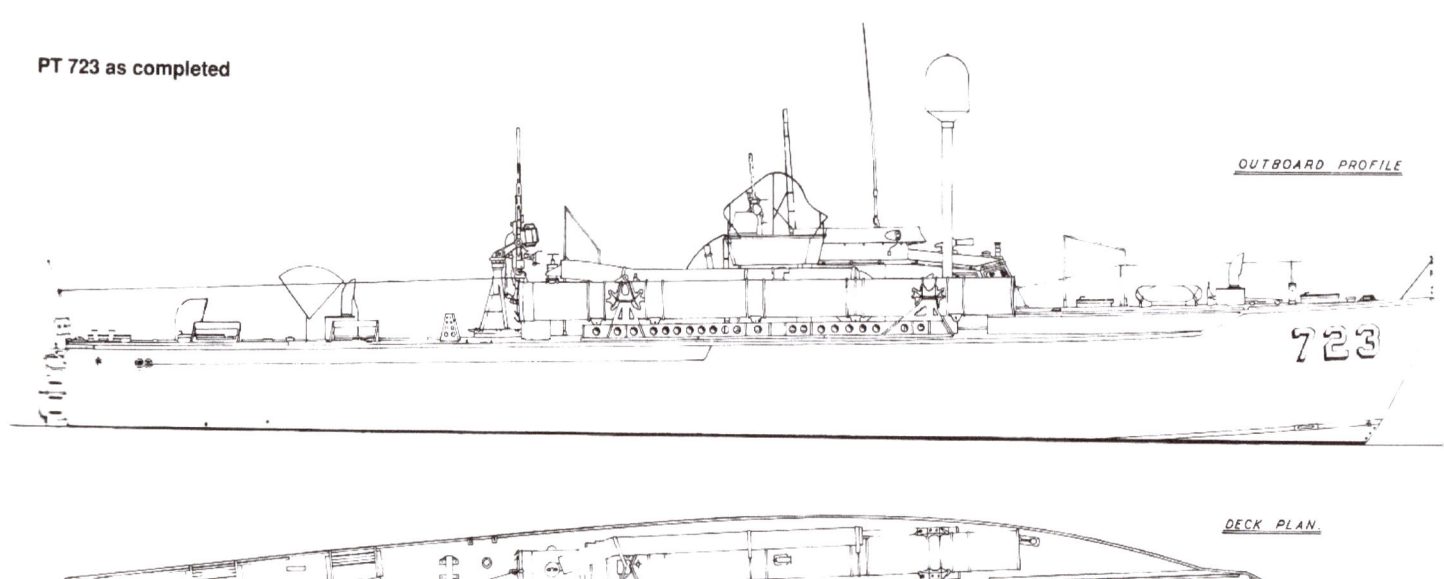

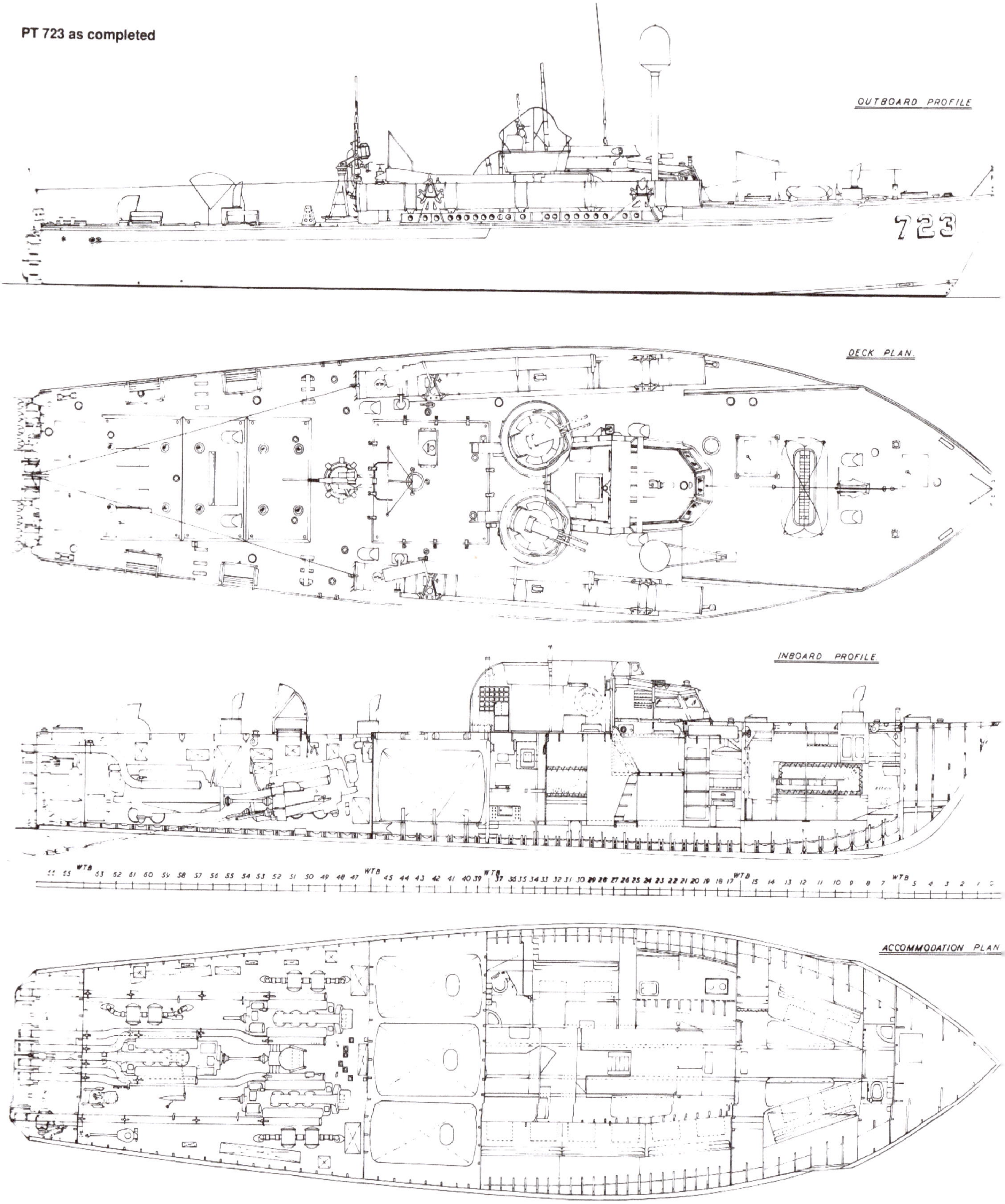

OUTBOARD PROFILE

DECK PLAN

INBOARD PROFILE

ACCOMMODATION PLAN

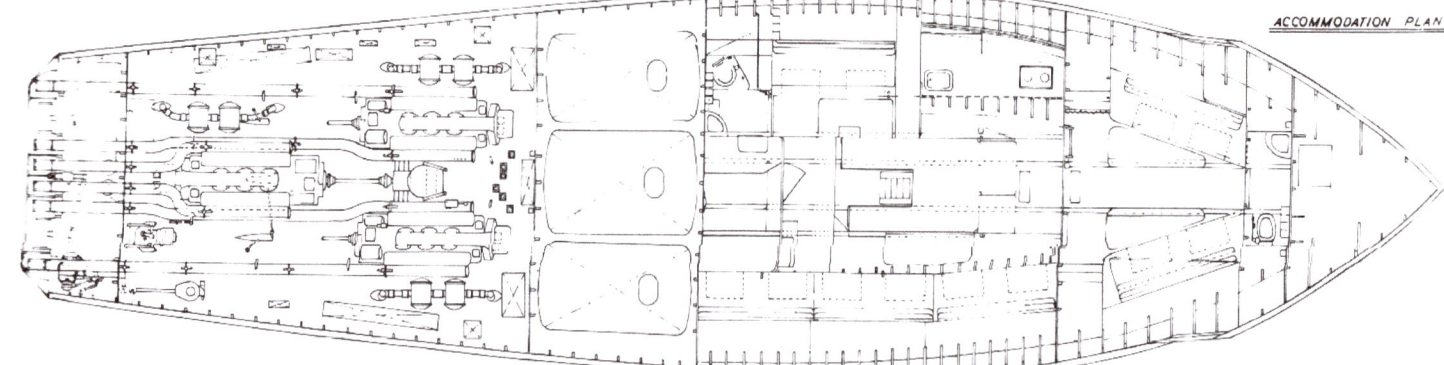

*General arrangement plan of a typical torpedo boat from the 2nd world war of the US Navy. Those boats where mainly used in the Pacific theatre.*

# Fast ships at all times

The world of naval warfare has changed a great deal since the end of World War II. The 1940s brought the disappearance of the battleship armed exclusively with heavy guns. It was replaced by the dominance of aircraft and submarines over the world's oceans.

Projectiles fired from gun barrels have been increasingly replaced by remotely guided or self-guiding missiles. Today, the combat ship capable of the greatest destruction of all is a very large, deep-diving, nuclear-powered submarine. It has been designed to never have to be used, if possible. The devastation that its more than one hundred thermonuclear warheads would cause is apocalyptic.

Very likely, the use of the weapons of one or two of these American 'Ohio' class submarines would spell the end of human civilization. Yet these ships would be several thousand nautical miles from their target, safe from any counterattack in the vastness of the ocean. Within only about 30 minutes of the launch order, their missiles would bring death, destruction, and centuries of the heaviest radioactive contamination.

Since the end of World War II, naval battles between ships have been very sporadic. The Israeli Yom Kippur War in 1967 brought the emergence of the guided missile fired against other ships.

Most often cited in modern naval warfare history is the 1982 Falklands War as a prime example of modern naval warfare. The use of nuclear submarines and guided missiles against surface ships, vertical-launch aircraft that operated carrier-based, and a modern landing operation with ground forces, were analysed in great detail using this post-1980 conflict and contributed significantly to the formation of modern naval strategies. Strategies that have already become obsolete and useless in today's world.

Today's adversaries in the so-called 'war on terror' have neither air fleets nor combat ships. Ship hijackings and attacks on ships in ports or at sea are the new threats to be faced. In addition, taking action against modern pirates is a new requirement for naval forces.

Today's naval soldier is increasingly given a hybrid role between military and police duties. Although commanders of today's naval units have addressed these challenges through numerous measures in the areas of training and tactics, naval armaments are very slow to come up with new equipment, weapons, and vehicles for asymmetric warfare at sea. Instead, submarine programs, the construction of ships specialized in anti-submarine warfare and the procurement of naval special continued.

In contrast, the navies of numerous nations are forced to enforce the new counter-terrorism measures with obsolete and partially depleted craft. As a result, many financial and human resources are being invested on maintaining this obsolete equipment.

The second Iraq war, which was certainly politically damaging and militarily largely pointless, now consumed the resources of the USA as a major power, which were urgently needed to transform its armed forces from the level of development of the Cold War to a 21st century force adapted to the new situation. In civilian shipping, new types of ships and designs have emerged in specialized areas whose performance far exceeds that of the equipment available for navies.

Today's large catamarans carry several hundred passengers and their cars, as well as numerous trucks, at speeds in excess of 30 knots, relatively independent of weather conditions. The advent of lightweight, high-powered diesel engines and the consistent use of lightweight construction methods has made these transport capabilities possible.

(Left) Symbolic of late 20th century weaponry, a frigate fires a standard missile.
Guided by RADAR, it targets an aircraft up to 20 kilometers away. This firing range could not be achieved by any World War II anti-aircraft gun.
The guided missle is the most important weapon of all navies today. It may be removed for LASER weapons in the nearer future. Also drones have got a significant importance in the sea warfare.

*(Above): This hovercraft, owned by the Navy of Greece, has lowered its vehicle ramp. The vehicle, which has a speed of more than 50 knots, can cross the entire Aegean Sea within a few hours.*

*It enables a rapid response in times of crisis to get support staffs or troops to the scene rapidly. In a nation, separated in several islands this is helpful to ensure security in all cases.*

The first navy to take advantage of the performance potential of large high-speed craft was that of the former Soviet Union. Spurred on by the development of the large ‚SR.N. 4' hovercraft from the UK, by the late 1980s the „Zubr" class (NATO designation: Pomornik class) had created an extremely fast generation of the largest hovercraft built to date at nearly 60 knots. These could drop several hundred soldiers and light combat vehicles on an enemy beach. These massive hovercrafts posed a major threat to the German and Danish Baltic coasts during the Cold War. The ‚Zubr' would have been unstoppable by the submarines and speedboats of the German Navy.

In a surprise attack, they could have quickly formed a bridgehead on the Baltic coast, which would then have been reinforced by heavy equipment a short time later. Such fast ships can only be engaged by fighter planes or helicopters if they can be brought into action quickly enough. Today, the Hellenic Navy maintains four newly built or second-hand hovercrafts of this type. They serve as a rapid intervention tool in case of occupation of an island near the Turkish coast.

Despite their mutual membership in NATO, relations between Greece and Turkey are still very cool. Parallel to the military function of the hovercrafts, they also help civil protection. It should not be forgotten that Greece is repeatedly hit by earthquakes. If, in the event of such a disaster, a section of the coast were to be blocked, these hovercrafts could deliver aid to nearby beaches in a very short time.

Meanwhile, the U.S. Navy has acquired an entire fleet of new types of high-speed ships the size of World War 2 destroyers. These ships are equipped with superior new technical „gadgets" and are highly automated. It will be seen if these ships can live up to the USN expectations placed on them. They should, because the expenditures to procure them were astronomical.

Another innovation is large fast ferry catamarans, which are in military service on the seas. Their task is the rapid transport of vehicles and tanks. Just as fast luxury liners ferried troops from America to Great Britain across the Atlantic during the World War, these high-speed transporters are intended to relieve the U.S. Air Force when transporting intervention troops.

No matter how fast a ship is on the water, an airplane will always be able to catch it, and today's seas are ruled from the air. Even for technically sophisticated warships, survival in combat at sea has become very difficult today. In the war over Ukraine, the sinking of two major warships of the Russian Black Sea Fleet clearly demonstrated the sensitivity of conventional surface units to quite inexpensive RADAR-guided missiles and to reconnaissance by drones.

The LCAC (Landing Craft Air Cushion) is one of the largest and most powerful hovercraft ever built. It was introduced by the U.S. Navy to land troops around the world on almost any beach. Unlike traditional landing craft, the LCAC does not have to worry about tides, currents, sandbars or muddy shorelines. It hovers over the water and can continue its journey at a remarkable speed even on firm and level ground. It is the main tool for the US Marine Corps to perform assault operations

*A LCAC maneuvers into the landing bay of an LSD (Landing Dock Ship). Such a ship often carries two LCAC that can land an entire Marine Corps combat regiment within a few hours. LSD usually carry the heavy equipment such as trucks, tanks, and guns of an expeditionary corps. The LCAC enables the Marine Corps to operate on the most beaches of the world.*

*A „Littorial Combat Ship" (LCS) built by Lookheed-Martin in the USA, is a ship specialized for combat in coastal waters. There, the 45-plus speed capability is very useful.*

*This is the look of one of the places where some of the fastest and largest aluminum ships are built: The facilities of Incat Tasmania in Hobart on the island of Tasmania.*

*The German Class 143 fast attack craft "Ocealot" lightens the scene with a shot of flares to protect itself against infrared guided missiles.*

# Speed at any price

*(Above) The American torpedo boat "PT-105" races across the Pacific Ocean. Torpedo boats never acted alone. Their strength lay in the group that, like a pride of lions, made a coordinated attack on a convoy.*

Since the invention of the torpedo in the 19th century, there have been small fast ships designed to attack and sink much larger and slower units. The natural target of a torpedo boat or destroyer from I. World War I was the battleship.

These massive floating fortresses were heavily armored and armed with heavy (28 to 38 cm caliber), medium (10 to 15.2 cm), and light artillery (3.7 to 7.6 cm). The guns of a small ship with calibers of 12.7 centimeters at best made these battleships unassailable.

Only a projectile with a large explosive charge - called a „torpedo" - hurtling along underwater in a straight line could do enough damage. The torpedoes of the time had only short ranges of several thousand meters.

The torpedo boat was therefore forced to venture at high speed into the firing range of the ship's guns and try its luck. Therefore, these boats were built very narrow and largely unarmored.

Whatever equipment was not absolutely necessary was not included. However, the small ships were inexpensive and easy to build. Once the technical secret of the torpedo, which had been made ready for series production by Robert Whitehead around 1864 and was propelled by compressed air, was learned, the boats now had a powerful offensive weapon against much stronger opponents. Torpedo boats gradually began to attract the interest of naval commanders from the time of the United States' Civil War onward.

The torpedo was the first effective way to blast large holes in armored ship hulls using the rather tepid explosives of the time. The gradually increasing armor strength together with the gun calibers made the battleships of the big navies seemingly invulnerable fortresses - a mistake, as it turned out later. The search was on for a weapon that could bring about a decision in a naval battle. a decision in a naval battle.

The first torpedo boats of the first hour were equipped with powerful and compact steam engines. The initially coal-fired boilers were kept running by stokers who had to slave away like madmen at ‚full speed ahead'. This is an interesting parallel to the ancient galleys with the oarsmen chained ...

Every new weapon sooner or later evokes a counter-weapon. The equally fast ‚torpedo boat destroyer' was invented to escort the capital ships like an escort fighter. It later became the destroyer as a universal type of warship. The torpedo boat, on the other hand, retained its role, although its appearance changed a lot.

The boats of the first World War had very slender hulls with mostly round frame cross-sections. They were therefore able to go fast mostly only because of their high propulsive power. Their seaworthiness was far behind anything else.

After the first World War, the large torpedo boat became less important as the destroyer was able to assume its duties as an

*(Above) This group of British MTB (Motor Torpedo Boat) returns from their „D-Day" mission in Normandy.*

offensive unit. On the other hand, the development of explosion engines, accelerated by the new aviation technology, called for a new small high-speed ship, the speedboat. It was basically very light in construction and followed an entirely different design approach from the old torpedo boat.

Most craft were gliders and could often reach speeds of nearly 45 knots. Their bulkhead cross-section was usually angular - called a „hard chine hull". They were made of hard plywood and were neither armored nor very durable. Even rifle bullets could penetrate the hulls. However, those very fast and maneuverable boats were built in large numbers, especially in Britain and Germany, and were incorporated into the tactical concepts of coastal warfare.

They were not particularly costly and could therefore be put into large numbers of them into service. In England, the fast boats were divided into the categories ‚Motor Torpedo Boat' (MTB) and ‚Motor Gun Boat' (MGB). The latter had only MGs and autocannons on board.

The Royal Navy fought intensively with its MTBs and other small craft in the English Channel area throughout World War II. True to British naval tradition, this was used to fight German convoys and fast boats as far as the estuaries of France and the Netherlands.

The crews of the MTB used a small vehicle of about 50 tons mass, armed with light weapons, two torpedoes and often depth charges. Powered by aircraft engines derived from those of the famous RAF-fighter Supermarine ‚Spitfire', aviation gasoline was carried as fuel. The service on board was correspondingly dangerous, often performed by young sailors and officers of the highest age of just 21. It was considered at least as dangerous as wartime service on submarines.

The German navy also benefited from aviation technology in propulsion matters. But this was the dying branch of airship aviation, which had born light and powerful diesel engines. Based on these engines, the German Lürssen shipyard was able to develop a new type of boat in the 1930s, whose hull had been optimized for the rough weather of the North Sea. With heavy armament, steadily improving armor, and diesel fuel that was not easily ignited, the German „S-Boot" was far superior to all other fast boats and therefore feared. Only Allied air superiority could curb the activities of the S-boat units.

The U.S. was slow to launch the speedboat operation in the Pacific. It was not until the American „PT boats" successfully held their own against Japanese naval forces in a number of engagements during the retreat from the Philippines that the deployment was intensified. Among other things, „PT-Boats" evacuated the famous General McArthur from the beleaguered Corregidor Island in Manila Bay.

The PT-Boats were used for reconnaissance, as drop craft of light landing units, for commando actions, and as a kind of „sea guerrilla." One of the most famous of these vessels was the „PT-109" which was under the command of John F. Kennedy.

Overall, the PT boats proved themselves quite well. They were able to use their high speed to their advantage during short „raids" that often lasted only a few hours at night. They were even able to defend themselves against air attacks by fighter planes. Changing course quickly or making full circles while firing all weapons simultaneously gave even agile fighters little chance of targeting the fast boats.

But their Achilles' heel was the gasoline that powered their engines. Even single hits from the tracer ammunition of an aircraft machine gun could lead to terrible conflagrations.

The German speedboats fought in all front areas, but the most intense battles were on the North Sea coast and in the Mediterranean. Tactics varied widely. In some engagements, the S-boats anchored themselves at buoys along the convoy routes with their engines shut down. Since they were now indistinguishable from the sea marks on the radar of the enemy escort vessels, the latter had no idea of the trap that awaited them. The sailors of the S-boats now listened for the propulsion sounds of the convoys and launched their attack at the right moment. The torpedoes fired by the speedboats at very short range rarely missed their targets and caused great devastation.

The battles were fast and extremely hard. Often they lasted only a few minutes. The so-called ‚hit and run' tactic was the core element of speed boat warfare. It was attempted to blanket the enemy with as many hits from machine guns as possible in a fast pass. A favourite tactic was to cross an enemy's course at

*(Above) The German „S-Boot" of later generations of construction was a feared adversary in 1943 due to its strong armament, the armor which is clearly visible here, and the seaworthiness of its slender hull.*

(Above) This already larger torpedo boat of the imperial navy around 1914 steams in formation with obviously high speed. Environmental protection was not yet an issue at that time, hence the black smoke from the funnels.

(Left) Sailors of the „USS Porter" (TB-6) engage targets with a light gun during the American-Spanish War in 1889. The 53.5 meter long boat was only 5.4 meters wide and could still reach 29 knots

over 30 knots and drop depth charges at the right moment. These were set to such a shallow depth that they detonated only a few seconds after being dropped. The powerful explosion could then tear the enemy boat to pieces. It is obvious that this tactic, approaching suicide attacks, required iron nerves. Not without reason, the young soldiers of both speedboat forces were regarded as young heroes with a short life expectancy..

The fast boats were made of plywood for the sake of lightness. The engine vibrations, wave impacts and numerous battle damages often caused the hull to age quickly. Many fast boats of this type were therefore replaced during the war. The German S-boats, built in a composite construction of diagonally laid plywood, were somewhat stronger and more durable. But even here, war-related losses ensured that only quite a few boats lived to see 1945. Those that did make it were handed over to foreign navies. In Germany, the advantages of this design had not been forgotten. The development was continued after rearmament in the German Federal Navy of the post-war area and thus created the basis for the introduction of extraordinarily successful speedboat classes in the following decades.

(Below) Sails may have been an option to improve the operational ranges of this steam engine driven early torpedo boat. It is unknown, if one admirality have taken this devices onto their shopping list in the early days of the steam driven fleets.

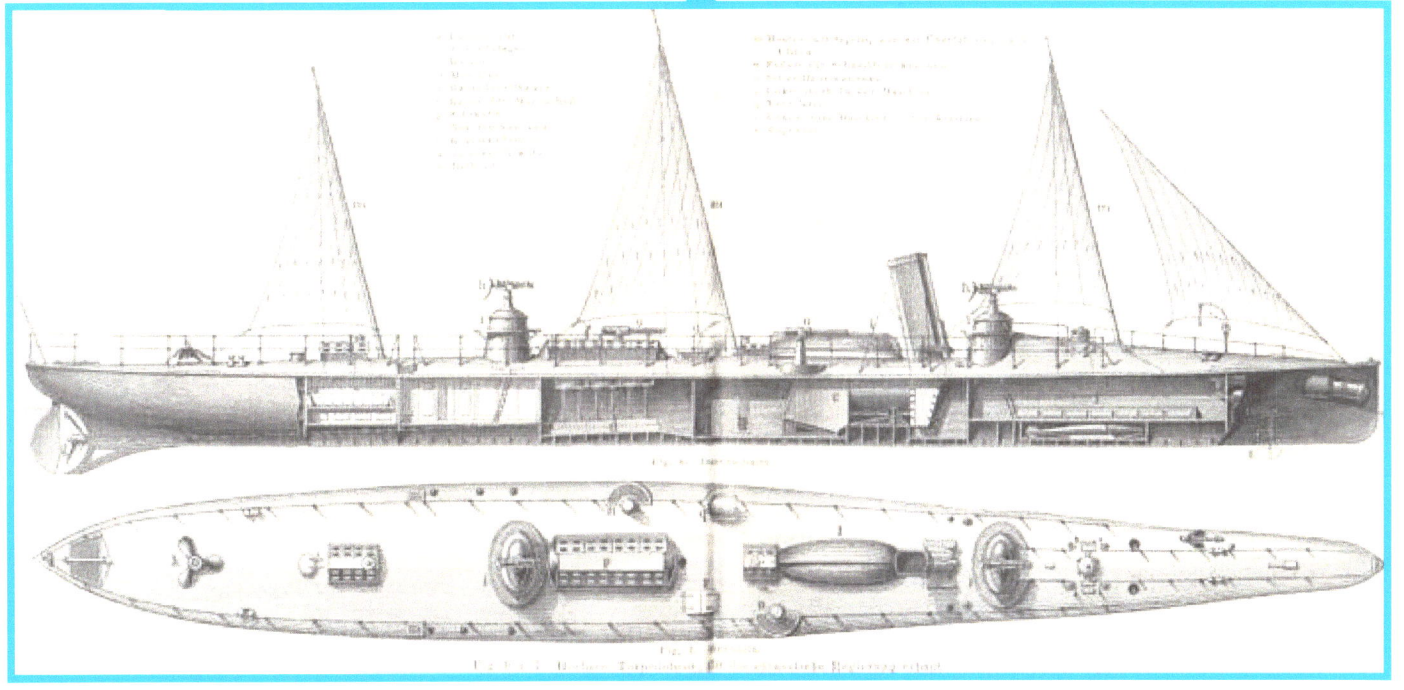

# The last of its kind

In many countries, shipping enthusiasts have often preserved treasures that are often very unique. The scrappings after the great wars have taken away many historic ships. But some have been preserved - including the torpedo boat ‚DRAZKI', built in 1900, which after all was in the service of the Bulgarian Navy continuously until 1950.

The ‚Drazki' is so unique because it is a torpedo boat powered by a single steam engine in the archetype of all fast boats. It is thus the only piece of this design still in existence. The Bulgarians have done quite a bit to restore it to its original state and are displaying it jacked up in the city of Varna. Of course, not all parts from the 1900s have been preserved, and so components from a sister ship were also used for the restoration.

A so-called triple-expansion steam engine of 1,900 horsepower served as propulsion. It propelled the boat to a speed of about 25 knots via a single large propeller. For a boat of that time this was a very good performance.

The boat was built by the Schneider et Cie shipyard in Chalon-sur-Saône, France. It was common for smaller navies at the time to stock up on ships and boats from the renowned shipyards of the major powers. Britain's, France's and Germany's shipyards made a lot of money from arms exports at the time. A parallel to today.

A product manufactured in large numbers by many shipyards were the then red-hot steam torpedo boats. They gave small forces the possibility of a more effective deterrent against large fleets, such as that of the Russian Empire in the case of the Black Sea, or against the navy of the Ottoman Empire. Torpedo boats were then a kind of „wild card" in the game against the „dreadnaughts" - the armored battleships that only economically strong nations could afford. Used in larger numbers, the torpedo boats posed a great threat even to heavily armored ships. Their torpedoes could hit the battleship underwater and cause a lot of damage in the right place with their great explosive power.

The „Drazki" was built completely as it corresponded to the typical requirements for torpedo boats in her time. Light guns and machine guns for defense against small boats and ships, but a heavy armament with rigidly mounted and a set of swiveling torpedo tubes. The crew was accommodated in a confined space. Most of the space had been reserved for the boilers and the steam engine.

The highly rounded hull and very narrow beam meant that such ships were not particularly stable in shape. The ship probably rolled quite a bit on the open sea.

*(Below) The torpedo boat was converted to a coastal patrol boat in the 1930ies. Surprisingly, the set of torpedo tubes amidships nevertheless remained intact, although their tactical value in the 30ies may have been doubtful.*

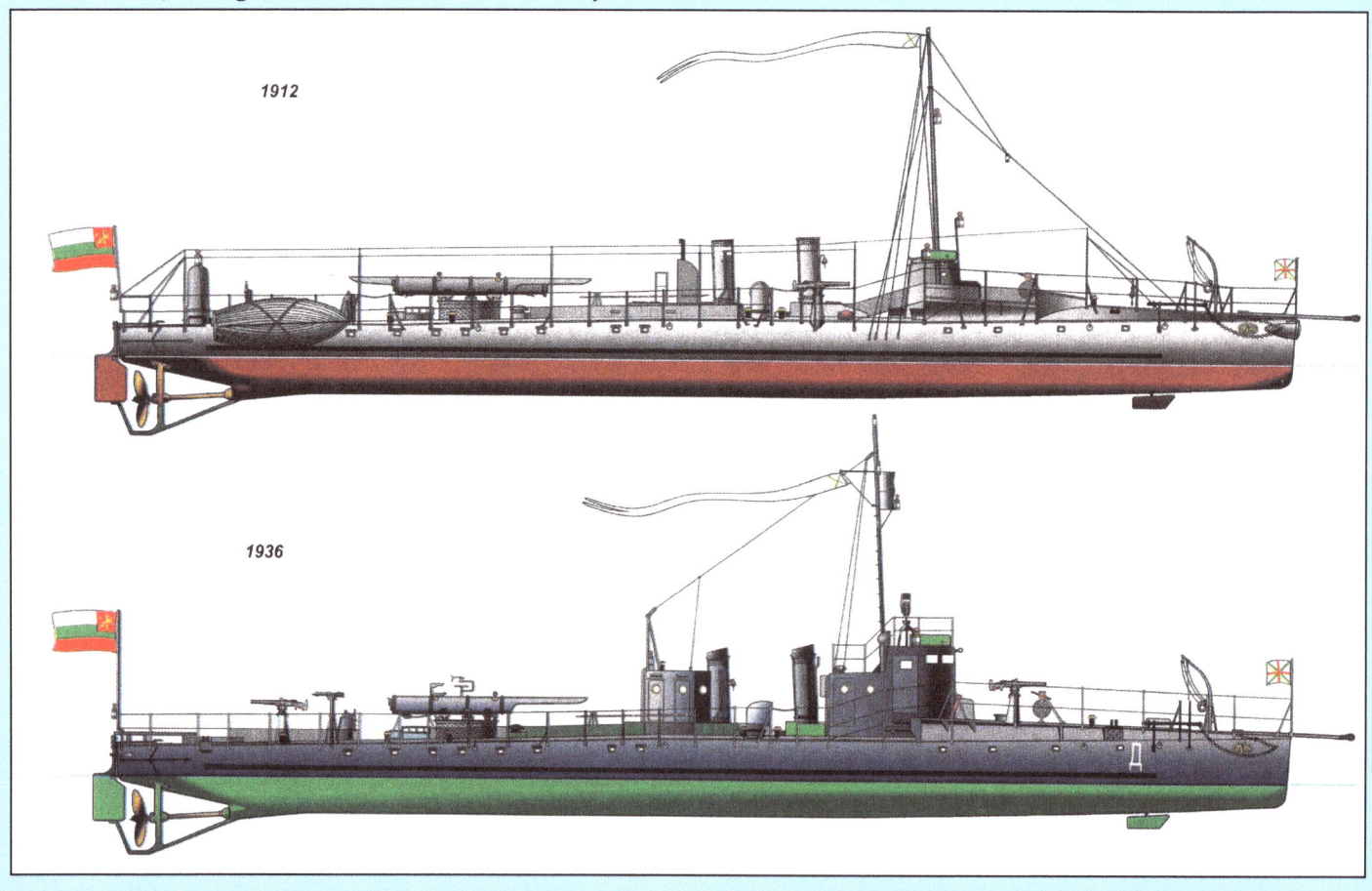

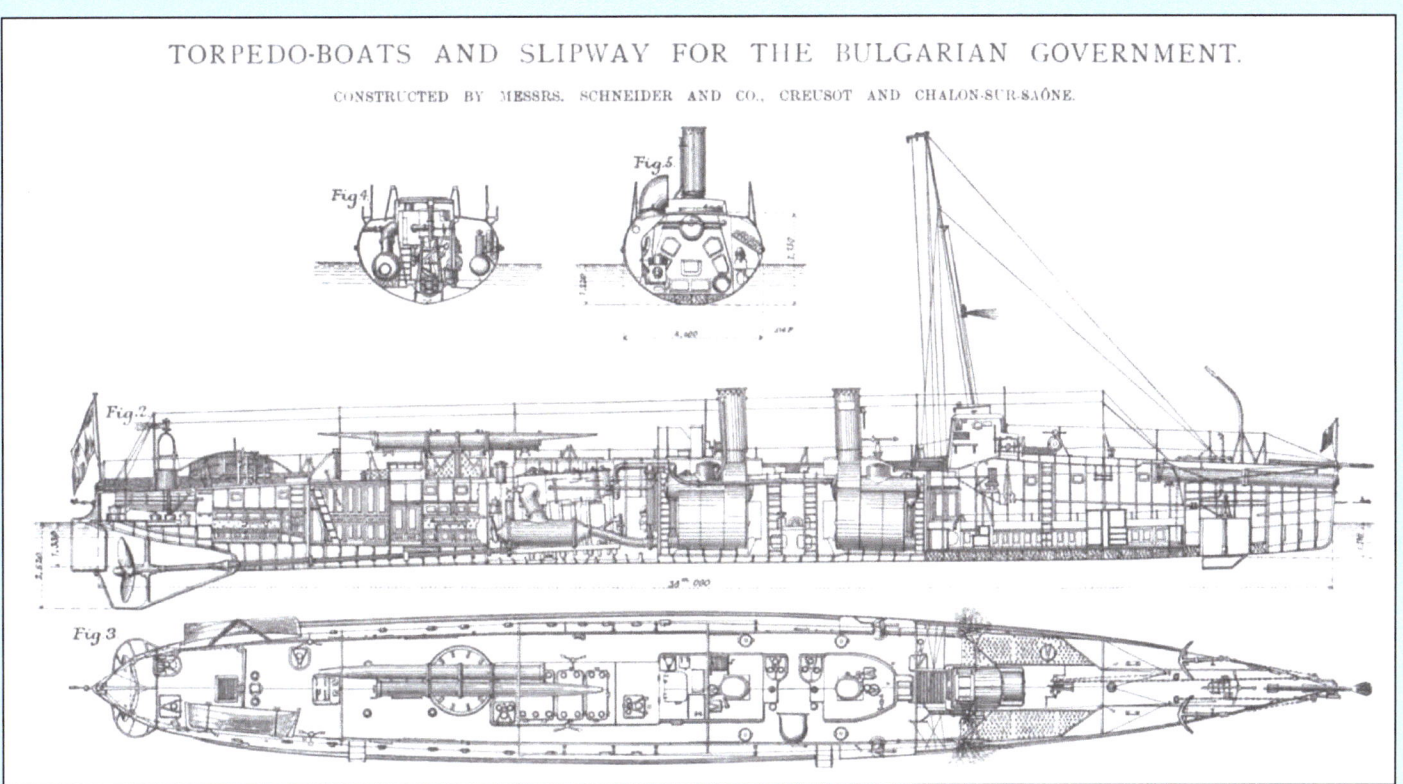

*(Bottom)* Side view of the type class of the „Drazki". These ships were bought from the product catalogs of the shipyards. Each basic design in catalog could then be slightly modified according to the customer's wishes. It is easy to see the large share of the propulsion system in the ship's hold. There was much less space for people. But the missions often lasted only a few days.

*(Right)* The „Drazki" on display in Varna. At the top of the bow, the single rigid bow torpedo tube can be seen.

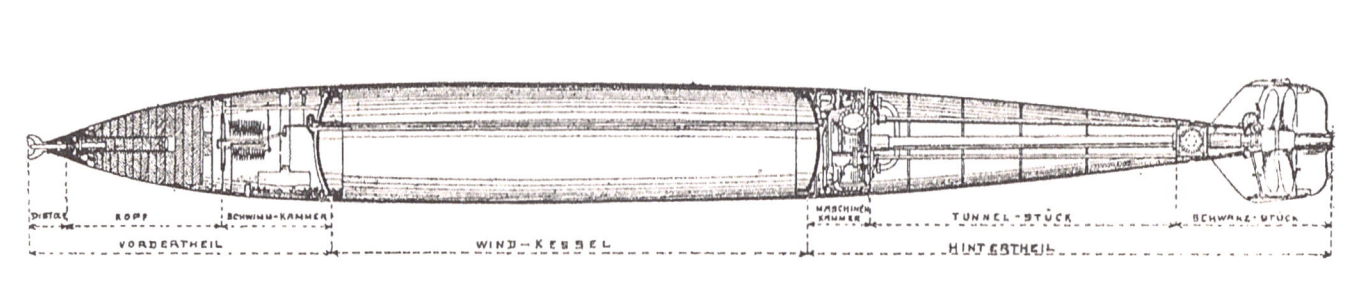

*(Above) The compressed air propelled torpedo perfected by Robert Whitehead in 1864 was a secret weapon for a long time. Wild espionage stories surrounded its later heart, the steering gyro and the counter-rotating twin-screw drive. Only the use of an apparatus similar to the gyrocompass for course stabilization and a water pressure gauge to maintain a set cruising depth turned the previously guide less torpedo into a very effective instrument of destruction. It also made it the first automatically guided weapon in history. Whitehead's first versions were initially held at a steady depth by a pendulum.*

*(Above) The „USS Winslow" (TB-5) was a sister ship of the „USS Porter". The picture shows the typical features of the first high-speed attack crafts of the late 19th century. The superstructure is rudimentary and armored to protect the crew from machine gun fire and splinters.*

*(Above) A picture of a 3.7 cm Gatling gun made by Hotchkiss (France) in the early days of the 20th century. She was a typical short distance weapon for small crafts to defend againts torpedo attacks.*

*The so called "Large Torpedoboat" (Here the S35) was the transition from the small slender steam torpedoboat to the first destroyers. The S35 was propulsed by turbines and reached 33 knots. the 79 m long ship was part of the Battle of Jutland and was sunken on the 31st May 1916 with all hands.*

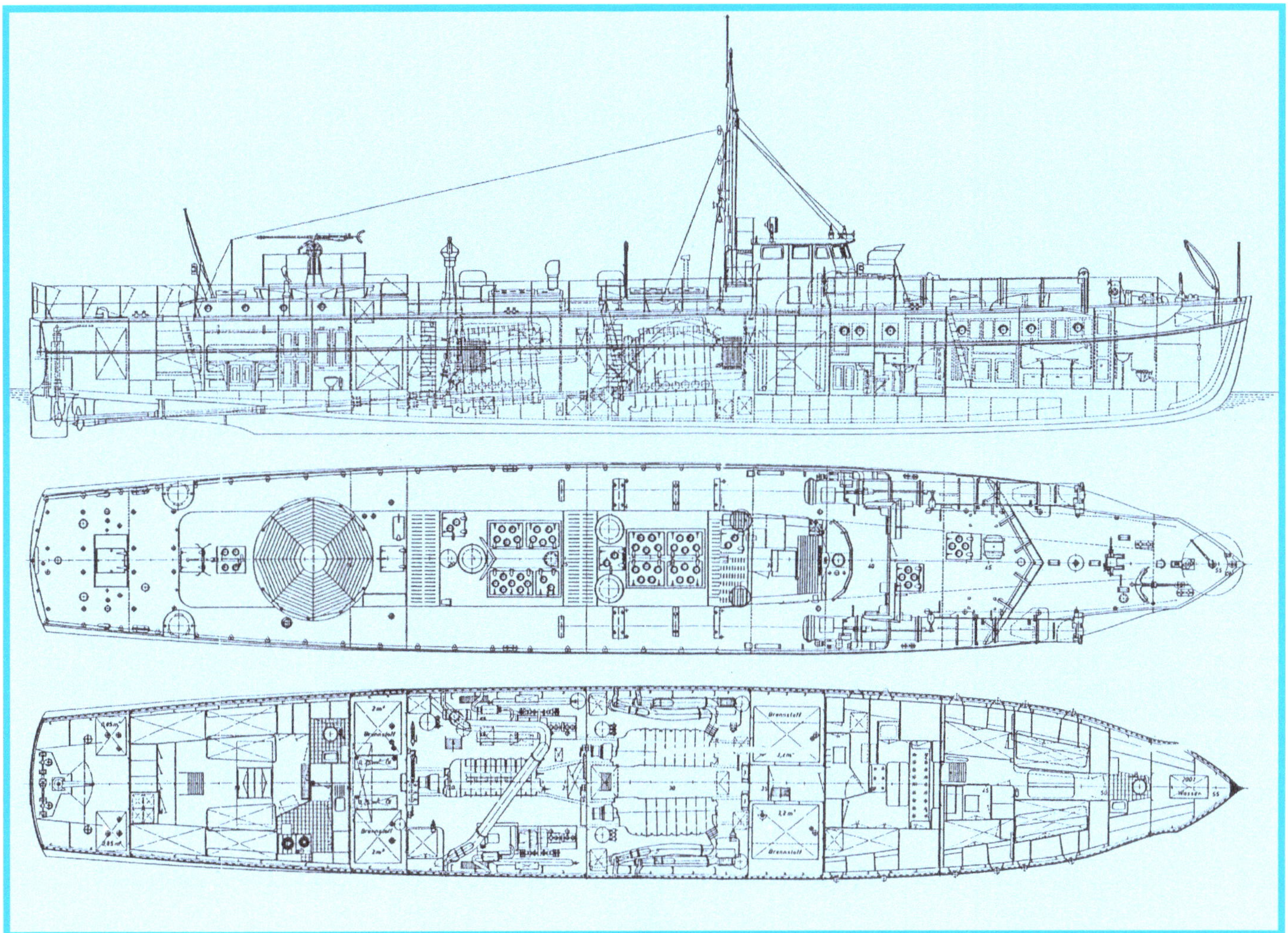

General arrangement plan of an German torpedo boat in the WW II. The S-7 to S-9 class appeared on the scene in the early 1930s. It still lacked the armor and heavy armament of the later S-boats, especially on the forecastle. But the hull and propulsion system were little changed.

Many features, such as the long slender hull with its three propellers and three angled rudder blades, invented by the Lürssen shipyard in Vegesack on the Unterweser near Bremen, have survived to today's generations of fast boats.

At 34 knots, the boats were not yet as fast as later models. Incidentally, instead of the fast-rotating airship engines, this example is powered by slower-rotating MAN diesels with seven cylinders each, which closely resembled the familiar submarine engines of the Germans. Since they vibrated strongly and were quite clunky, they did not catch on in the S-boats. The side section shows the construction of the hull with wooden frames and a wooden keel. Multi-layered plywood served as planking.

On the English Channel and in the North Sea, the German E-boat and the Fairmile-C Motor Gunboat (MGB) were natural opponents. While torpedo boats mainly attacked larger enemy ships, the MGBs mainly fought small armed units such as armed trawlers and especially the E-boats. MGB were equipped with machine guns and -canons and depth charges for short range combat.

*(Above) In the German postwar Navy, cooperation between air and naval forces has always been essential - something that sets it far apart from the old 'Kriegsmarine', which had consistently failed at this. The Westland ‚Seeking' helicopter peeks over the horizon with its radar, and the fast boats then fire on the target it has spotted. The helicopter is barely visible to the enemy, as it can very quickly duck back behind the horizon. Communication between the two took place only through automated data exchange, and is thus almost impossible to detect. With this combination, the small vehicles of the German Navy were a very serious deterrent force. Among the many missile speedboats built during the Cold War period, the German Class 143 is considered one of the best. Surprisingly, the hulls were still built of multi-layered plywood - but in a modern way. However, Lürssen shipyard knew how to handle the material wood due to many years of experience. The result was extremely durable yet lightweight hulls*

*(Below) Launch of a 143-Class fast attack craft (FAC) at Luerssens Shipyard near Bremen (Germany) in the early 70ies..*

# Fast combat boats after the war

The fast combat boat has not been equally well regarded in every navy. The United States Navy has had little use for these small craft since World War II. The U.S. fleet has been consistently shaped around the new giant aircraft carriers and nuclear submarines.

The short range of the fast boats made them unsuitable to keep pace with the fast aircraft carrier formations. Therefore, they were of interest to the U.S. Navy only when conducting long-term operations along the coast of a war zone. However, for controlling coastal freighters and junks on the coasts of Vietnam, small fast boats were better suited than destroyers and guided missile cruisers.

The Americans therefore introduced the „Ashville" class in 1966. These boats, about 51 meters long, were quite fast at 35 knots, but armed only with a few machine guns. Their main function was to act as police craft off Vietnam. But they had little in common with the high-powered combat ships that the navies of the Soviet Union and many Western countries acquired.

To this day, the high speed and manoeuvrability of the speedboats is bought at the price of a short range at full speed. Therefore, the nearest refuelling station should never be too far away. The most suitable coastline for speedboats is one that is rugged with bays, fjords and islands. With proper preparation, it offers numerous hiding places from enemy aircraft and from submarines. In many navies, the supply vessels of the speedboats are often not very large ships either, so that they can operate quickly and flexibly.

In the German Federal Navy, for example, the Type 401 tender (Rhine class) was heavily enough armed with two 100 mm guns and numerous anti-aircraft weapons to be able to act as a serious opponent to lighter combat ships.

Speedboats are the guerrillas of the seas. No navy has ever lived out this concept as intensively as that of the Kingdom of Sweden. Hideouts and even bunkers were built in the archipelagos and bays of Sweden to provide the small light combat craft with a safe hiding place from an overpowering enemy.

At night, the speedboats would come out of their hiding places and go hunting. Even if an enemy had partially occupied the Swedish mainland, the fast boats of the Spica and other classes operating from hiding places would have repeatedly struck unexpectedly from ambush, making it difficult for the enemy to supply at sea. After World War II, Germany was also intensively rearmed in the wake of the formation of NATO. The U.S. and England donated destroyers and other craft, but when it came to fast boats, the experienced commanders of the early 'Bundesmarine', all of whom had still fought actively in the Great War, preferred to rely on home grown products.

Thus, the first thirty boats of the „Jaguar" and „Zobel" classes laid the foundation for modern equipped fast attack units. The German designs continued the principles learned with the S-boats. Long slender hulls, light and powerful diesel engines, and composite-layered sideboards made them the most suitable craft of their type for the North and Baltic Seas at the time. The new German S-boats were armed throughout with two 40 mm Bofors guns and initially unguided torpedoes. Later, they were converted to wire-guided torpedoes.

The new boats were extremely agile and fast, even in higher sea states. Skilled helmsmen could steer them so that they surfed down long wave crests and were not exposed to the strong blows of the waves. It must have given the crews a special feeling of exhilaration to sail these boats fast. The next generation of fast boats then set completely new standards worldwide.

At Lürssen shipyard in Vegesack near Bremen, much larger new boats were built, the Class 143 and 148, whose armament now surpassed the firepower of a wartime cruiser in certain respects. The artillery component consisted of the Italian Oto Melara fully automatic 76 mm gun. It could engage both sea

(Left) An American attack craft of the „Ashville" class. Unlike the FAC of Europe, it has been specially equipped for warfare against guerrilla forces and for police duties. Torpedoes and guided weapons do not play a role. The „Ashville" boat is therefore to be regarded more as a gunboat.

and air targets very effectively. The smaller 148-class boats got a 40 mm Bofors on board at the stern. The larger Albatros class received a second 76 mm gun instead.

But the real main armament was the new Exocet missile. This low-flying missile could carry a fragmentation charge indigestible to ships about 70 kilometers to a sea target. Since this missile covers the distance in blind flight without radar emission, it often cannot be detected in time by the enemy. Only shortly before the presumed location of the target does it switch on its radar seeker. The effect on destroyer-sized ships is lethal. The damage of a single hit with an Exocet missile will incapacitate any of these battleships. The Exocet MM38 can only be defended against by fully automatic guns or special missiles.

It has already landed several hits in the course of its „career": The British destroyer ‚Sheffield' and the transport ship ‚Atlantic Conveyor' sank due to the impact of a missile hit. In the Persian Gulf, the US frigate ‚Stark' was very badly damaged. If only a single missile of this type is already so dangerous, the effect of a salvo of several 143-class fast boats can only be considered devastating.

Therefore, the fast-boat flotillas operating from ports on the German North Sea and Baltic coasts quickly became an important element of deterrence. In some NATO maneuvers, the German Navy's fast-boat flotillas were able to attack entire U.S. Navy aircraft carrier formations very successfully in a simulated manner, without the latter having any prior idea of the fast boats' presence.

The new boats with long slender hulls spread all over the world. These boats are ideal for small navies and armed forces that have to perform police duties in addition to a defensive role. They can be equipped as lightly armed patrol boats, as missile-firing combat ships, or even as carriers for smaller helicopters in anti-submarine warfare.

Armament ranges from light machine guns to light automatic guns up to 76 mm caliber, and sea- or air-launched missiles. The German 143-class fast patrol boats, for example, were retrofitted in the 1990s with the RAM missile for close-range air defense.

But the seaworthiness of these boats is still limited - as is their range. So it's no wonder that the fast boats are operated as watch-free vessels in many navies. Everyone on board remains at their station from the time they set sail until they return. After all, such a deployment can only last up to 48 hours at combat speeds of between 28 and 40 knots, since after that the fuel will have been used up. Even today, operating on fast boats is dangerous and challenging. In addition to the cramped quarters on board, the often abrupt ship movements and the considerable engine noise during high-speed travel, there is also the nervous strain of combat operations, which can become very great even during maneuvers.

The enemies of the speedboat can be found in all sorts of places: Submarines can target them as well as fighter planes or destroyers. The speedboat cannot rely on particularly strong defensive armament or armor for protection. Only the speed, the aggressiveness of its deployment and the surprise effect are on its side.

(Above) At the beginning of the German Federal Republic's rearmament, which was not without controversy in the 1950s, the new 'Bundesmarine' (German Federal Navy) had to resort to the material that could be procured most quickly. Boat P-6192 (below) belonged to the Norwegian „Nasty" class. The boat P-6053 belongs to the „Silver Gull" class and was built by Lürssen according to plans from the World War. The boat above is one of the „Zobel" class and a new development according to the requirements of the German Fleet.

Of course, no modern speedboat squadron operates entirely on its own. The naval command dispatches formations of several boats for precisely planned ambushes or attack manoeuvres based on carefully evaluated reconnaissance data from the enemy's electronic emissions, radar information from its own air reconnaissance, or even satellite information. A modern computerized fleet command and control system transmits the data to the flotillas, which can then approach the enemy undetected even without using their own radar. In the German Navy, a fleet command and control system was introduced as early as the 1970s, providing the fast patrol boats with accurate launch data for the Exocet missiles.

During the 1973 Yom Kippur War, Israel Navy boats engaged in various situations with their Egyptian counterparts, which were from the Soviet Union. Not only were missiles fired, but the guns were also used to destroy the enemy. It became apparent that the manoeuvrability of this type of ship and its acceleration capability were essential factors for survival.

In the various clashes between U.S. Navy aircraft and Libyan Navy units in 1986, on the other hand, it became apparent that fast patrol boats without adequate missile armament or aerial cover do not stand a chance against an enemy that has air superiority. Two modern speedboats were taken out by air attacks, while a third was neutralized by the U.S. cruiser ‚USS Yorktown' through a Harpoon attack. Air defense was always the biggest shortcoming of the fast boats - no matter to which technical generation they belonged.

The small size of the ships and the simultaneous need to equip them with the strongest possible offensive armament mostly reduced the boats' defenses to barrel weapons. It is only since the 1990s that Western fast boats have been equipped with anti-aircraft missiles such as the American „Stinger" or the French „Mistral". But since these have to be aimed and fired by hand, this is only a makeshift solution.

Other navies, such as the Israeli Navy, have retrofitted Gatling guns of the „Vulcan" type. These fully automated weapon systems are compact, very firepowerful and extremely responsive. They can even shoot down incoming artillery shells. But the range is very short, making them a „last minute" weapon. Various development programs in the West have resulted in light radar-guided missiles that, like their larger sisters on destroyers and cruisers, are all-weather capable and have a range of several kilometres. But because mostly fast-attack craft are of secondary importance in nations' budget policies, few ships have been equipped with them to date. Moreover, the performance of such light weapon systems has been repeatedly challenged in discussions. Overall, a new set of interests for the navies of the West and East developed after the end of the Cold War.

The navies of Russia suddenly received only a minimum of the resources that had been available during the Soviet period and had to fight for their survival. They could therefore focus only on maintaining their most important units, such as the missile cruisers and the large nuclear submarines.

Western navies were given entirely new roles as they became more and more involved in UN missions. Hunting submarines in the North Atlantic and keeping landing craft away from Western European coasts was no longer a core objective. There were widespread budget cuts, and again, fast patrol boats were one of the main casualties.

The trend was to replace these craft with new types of slightly larger combat ships, which were classified as „corvettes. "Corvettes in World War II were vehicles built exclusively to fight submarines. They were not very fast and had very specialized armament. Depth charges, anti-submarine mortars, and light guns were more important than anti-aircraft guns or anti-ship attack weapons. They were often more seaworthy in design and had a greater cruising range than the destroyers of their day. The use of the British „Flower" class corvettes and their slightly larger successors in the years between 1940 and 1945 was critical to the war effort.

After the war, the tasks of the corvettes were expanded. A distinction was now made between defensive operating ships such as the frigates and offensive destroyers. The corvettes were now too small to carry helicopters and anti-aircraft missiles in addition to anti-submarine warfare equipment. And the fast boats were also getting bigger and stronger. Corvettes were now built only as low-cost units for secondary missions. Best known is the D'Estienne-D'Orves class, which France built in larger numbers in the 1960s.

But in today's world, warships that can be deployed at low cost and are not expected to reach extreme top speeds are needed. Today's weapon systems are more flexible and compact than they were twenty years ago. Thus, it is easily possible to accommodate a well-sorted mix of weapons and equipment on a fairly small hull. By choosing a speed of between 25 and 28 knots as the maximum, the propulsion system could be kept compact and economical. The range of these new corvettes is much greater than that of the fast boats.

Corvettes of this type represent a viable solution when it comes to guarding coastal sections to protect fishing rights, fighting crime, or shielding coastal sections in combat zones. They can defend themselves sufficiently well and are flexible. They usually have helicopters on board, which allow boarding or surveillance of other ships even at further distances. The new corvette has thus become a universal ship.

It seems that the speedboat has had its day. In fact, hardly any speedboats are ordered by navies anymore. The focus is on new types of ships that may define the future of naval affairs - superfast troop carriers and STEALTH ships. But more of that later.

# A case study: speedboat war in Vietnam

(Above) A type of boat made famous by the movie 'Apocalypse Now', this PBR Mk II races through the Mekong Delta. Called 'Pibber', these boats were especially suited for the smaller arms of the Mekong River due to their jet propulsion and shallow draft. They were armed with a forward double MG M2 (Cal. 50) and an MG M60 (7.62 mm) aft. In addition, there were 40 mm grenade guns and the M16 rifles of the 3 to 4 man crew.

The Vietnam War was not only a new and traumatic experience for U.S. ground forces, but the U.S. Navy also had to adapt to a type of warfare for which neither its equipment nor its strategies were suited. Instead of naval battles fought with missiles and aircraft, or duels between nuclear submarines taking place in the depths of the seas, they suddenly had to bomb small and barely detectable targets in the jungle. There were also new tasks on the coast: Checks against the smuggling of weapons into South Vietnam and the surveillance of rivers and coasts was the new battleground of this war - a task neither missile cruisers nor submarines could do. So, in the mid-1960s, a new breed of small speedboats were created, specially equipped for the new kind of war. They embarked on the often extremely dangerous patrols into the Mekong Delta and the obscure coastal sections of South Vietnam. In addition to the unbearable hot and humid jungle climate and mosquitoes, the missions brought strains from constant skirmishes with Viet Cong forces or small ships smuggling weapons from North Vietnam. While a speedboat on a river is a highly visible target, the boat crews often could not even see from where they were being fired upon. Often, with considerable waste of ammunition, they simply 'chased steel into the heath'. Whether one hit anything in the process, no one knew.

But it was not only from the Viet Cong that danger threatened. Sandbanks, mines and invisible underwater obstacles such as tree trunks were a constant threat. The U.S. Army had been using Agent Orange, the notorious defoliant containing the highly toxic substance dioxin, since 1963. The poison, which was sprayed over large areas from airplanes, also damaged many U.S. soldiers, who often became ill with cancer as a result years later. This also affected riverboat crews, who often had to sail through areas where 'Agent Orange' had been sprayed.

The war on the rivers of Vietnam was one of the most dangerous operations of the war. When it became clear that the fast boats were too vulnerable, they went back to the 19th century and reinvented the river monitor: an armoured, slow, but also heavily armed motor propulsed barge.

(Below) In the Mekong Delta, U.S. forces deployed three Bell SK5 hovercraft in the late 1960s. These were licensed from the British Hovercraft Cooperation for the SR.N5 "Warden Class" type. The introduction of the hovercraft was not as successful as hoped. The vehicles were not very effective because they were not armored. Also the skirts were too vulnerable in the terrain.

(Above) A group of ‚Swift' PFC escorting slower crafts. This type of boat was used in large numbers on rivers and along the coast of Vietnam

(Below) Vietnam War-era river combat boats erected at Naval Amphibious Base Coronado (California) to commemorate the fallen: on the left, the „Pibber"; in the center, the „Swift"; and on the right, a heavily armed river monitor that was slow and armoured. In South Vietnam, a variant of the „Swift" was even mass-produced from concrete.

*(Above) A German post war destroyer fires a practice shot with an MM-38 Exocet.*

*(Below) Taken out of the wartime this cutaway drawing shows the inside of a American PT-boat of WW II*

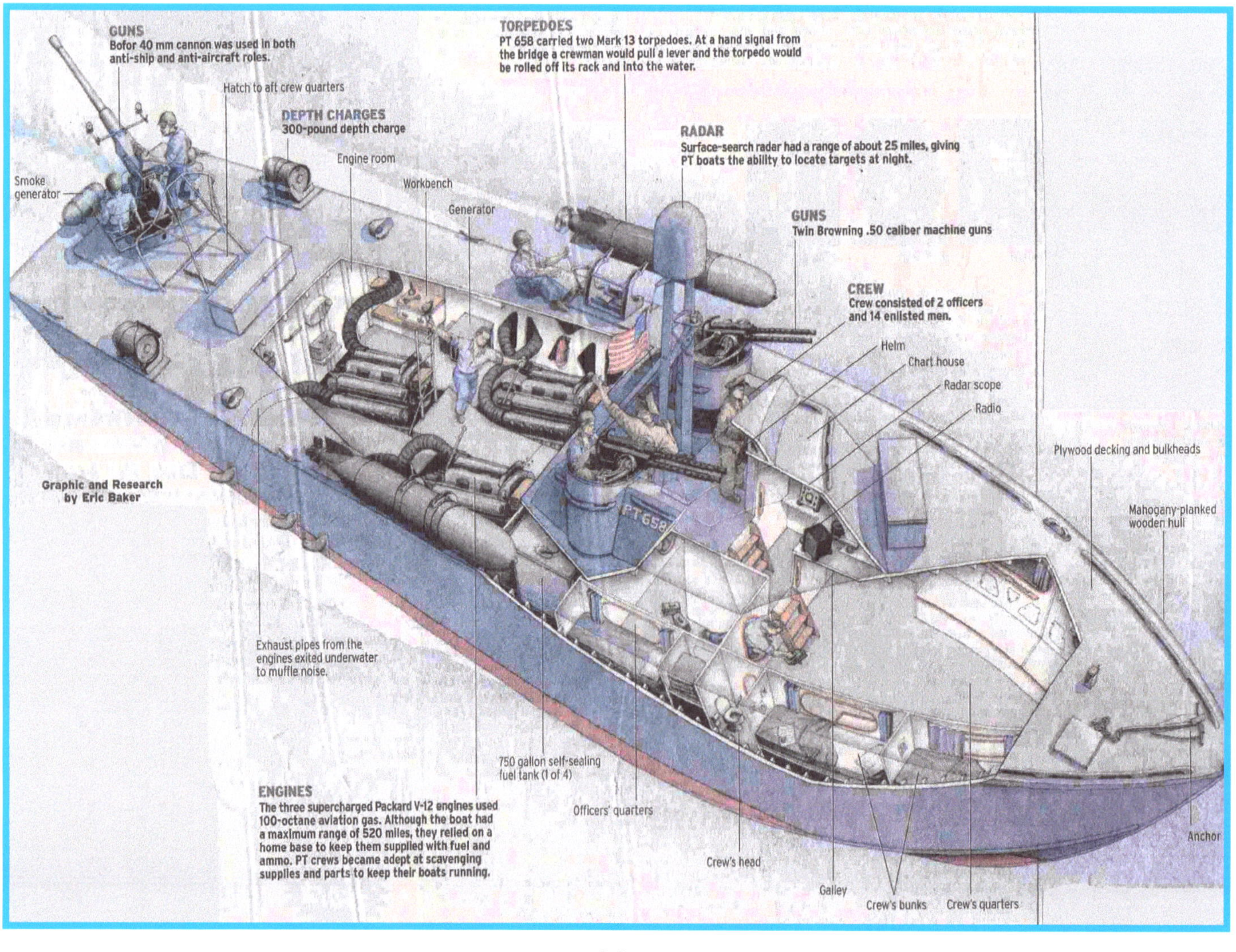

*(Above) The Italian perennial favorite: an Oto Melara 76 mm universal gun aboard a U.S. Coast Guard ship. The lightweight and compact gun system has become standard equipment on many navies.*

*A Germany 143 class fast attack boat cruises the course of its predecessor at more than 30 knots. Formation maneuvering has been developed into an art in the speedboat branch. Coordinated maneuvers within a group are part of attack tactics and improve each boat's chances of survival. This skill constitutes part of the self-confidence of speedboat drivers.*

# A true ship killer: the Exocet missile

(Left) The fist of a modern warship: an Exocet MM40 takes off: The MM40 is a modernized current version of the MM38, which came to public attention after the 1982 Falklands War. A single missile shot can hit a frigate or destroyer so hard that they are no longer operational. Thus, an older fighter such as the Argentine ‚Super Entard' was capable of sinking HMS „Sheffield" from a distance of about 25 kilometers.

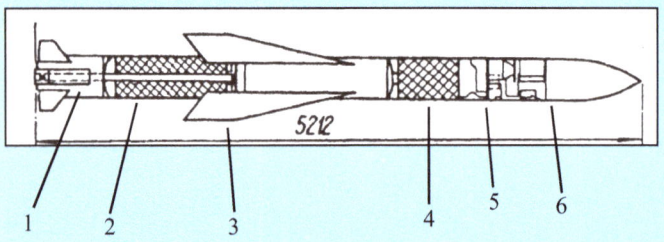

(Left) A small view inside an Exocet MM38:

1 launch rocket
2 Rocket motor for cruise flight
3 Wings
4 Splitter warhead
5 Autopilot, inertial navigation system and radar-Altimeter
6 Homing radar

(Right) The flight profile of an exocet:

1 Launch
2 Ascent to 100 - 250 feet
3 Cruise flight at Mach 0.93 without radar at low altitude
4 Active target approach (radar use)
5 Final approach after target acquisition
6 Impact just above waterline
7 Course correction after target acquisition

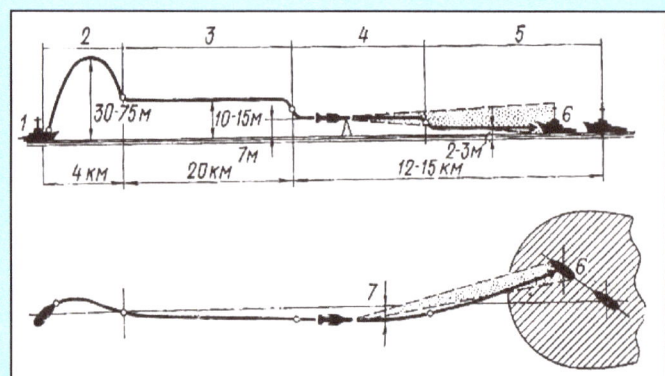

The Exocet operates according to the same scheme as a kamikaze aircraft in World War II: it heads directly for the target as fast as possible and has no fear for its own life. Because this class of missile performs a cruise at very low altitude and high speed without using radar, it is difficult for the targeted ship to detect.

Although the warhead is small compared to an artillery shell of large warship, it destroys electrical and other equipment on board by splinters. The rest of the propellant ignites combustible materials such as the aluminum of the superstructure. This is what happened aboard the sunken HMS Sheffield.

Modern warships are not as heavily armored as the battleships of yesteryear. Aluminum is a particular source of danger, as it easily catches fire. The only chance of escaping a hit by this missile is a sophisticated missile defense system such as the British Seawolf, the U.S. RAM and the Goalkeeper gun system (CWIS). Beeing stealthy is also an option.

Other guided anti-ship missiles also operate on the same principle as the French Exocet. They often can be used on ships, a large number of aircraft, and also from trucks on the coast. The firepower of a B-52H bomber, for example, is enormous, as it can be armed with 12 Harpoon missiles each. Just a few such bombers can wipe out any surface fleet from a safe distance of about 200 kilometers.

A salvo of 16 missiles fired for example by four fast attack boats would have been absolutely devestating to a Soviet assault fleet in the Baltic Sea. The Excocet gave small ships the firepower of aa heavy cruiser of the WW II. For many Navies the French missile is the standard surface-to-surface weapon.

(Left) A French-built ‚Super Entard'. A not very impressive aircraft compared to other carrier based fighters airplanes, but a major threat to the British Expeditionary Force in 1982.
The air attack on HMS Sheffield was launched from the only Argentine aircraft carrier. Therefore, the later sinking of the cruiser „General Belgrano" by a British submarine was a hard blow for the Argentine Navy.
This forced them to withdraw their aircraft carrier from the game. Without this measure, the losses of the British forces might have been even greater.

# Form theory and driving forces

Even in the pre-war period, naval engineers made numerous efforts to significantly increase the speed of ships. The most advanced at the beginning of World War II were certainly the German shipbuilders and engineers of Schärtel and Sachsenberg.

The research area of both was in the field of hydrofoils. They were able to establish a technological leadership position by the end of the war, which enabled them to successfully use the results of their work in post-war civilian shipping.

Any conventional ship gains its buoyancy from the balance between its own mass and the mass of the water it displaces. For this reason, conventional ships are called „displacement ships." This principle is fundamental and applies to any floating object - whether it is a paper boat or a supertanker.

When such a body is moved through the water by the force of a propulsion system, the water will avoid it as long as it can. Two obstacles stand in the way of the floating body:

The first is pure water resistance, which increases as the square of the speed attained. A rower who wants to propel a rowing boat must therefore row not twice but four times as hard if he wants to reach 4 km/h instead of 2 km/h. The second obstacle is the bow wave.

The second obstacle is the bow wave that builds up in front of the boat. It consists of water that has chosen the upward path instead of swerving to the side, thus piling up as a mountain in front of the bow. Since the ability of water to build up mountains above the surface is limited, it eventually flows out of the jam and thus causes the bow wave.

Behind the ship, the displaced water flows together again. There, the inertia of the water masses makes them collide so strongly that a more or less high „stern sea" is formed there.

In addition, the water flows along under the hull and pushes upwards at the stern into the „gap" left by the ship.

The extent to which these influences affect the hull of the ship is determined by the propulsive energy expended and the shape of the hull. A long slender hull with a sharp bow creates less water resistance than a wide blunt hull. It can also more easily split the bow wave and push it to the sides.

But pushing the water aside and splitting the bow wave has its natural limit. The force of the water masses flowing against the bow increases with increasing speed to such an extent that the hull of the displacer no longer has sufficient inertia to continue to divide it. This limit can be determined by calculating the hull speed, a function in which the length of the waterline plays a decisive role.

If a ship is driven to exceed the hull speed by a propulsion system of great efficiency, the ship's condition must inevitably change. It can no longer push the water aside, but begins to lift its bow out of the water as the bow wave increasingly pushes under the hull. If the ship has not been designed accordingly, it may capsize. But an appropriately shaped hull will begin to glide.

This is the moment when buoyancy is no longer generated by the weight of the water, but by the pressure exerted on the hull by the water rushing underneath it.under the water exerts on the hull. Such a vehicle is called a „glider." There are also intermediate states in which a vessel is already gliding at the forecastle, but the stern still floats as a displacer. These are called „semi-gliders."

The planing behavior can be controlled by a special shaping of the hull. The most primitive of all gliders is the surfboard

*(Below) This model hull is towed in a ttest tank to research its behaviour when running at high speed. Tank model test are required to testify the computer calculations with CFD (Computed fluid dynamics) software.*

shaped, which relies only on a straight and smooth bottom. Then there is the classic V-shape and various intermediate shapes that belong to the semi-gliders. In addition to the appropriate hull shape, gliding also requires a low mass and a strong drive.

If the boat becomes too heavy, it can no longer be carried by the pressure of the water rushing under it. Therefore, most gliders are small craft and the large fast craft are usually semi-gliders.

The seaworthiness of the gliders and semi-gliders quickly reaches its limits when wave heights rise well above 1.5 meters. The vertical acceleration generated when entering wave troughs and jumping over wave crests become dangerous for the ship and crew in higher sea states. Especially in the forecastle, a crew member can quickly be thrown off his or her feet.

Another dissatisfaction of the speedboat operators concerned speed. The 30 to 40 knots achievable by conventional means were no longer sufficient after the war. New types of gas turbine drives were tried out, or, as in the Soviet Union, compact radial engines were tested as propulsion systems to cram even more power into the light boats.

However, technical limits were quickly reached because of the glider hulls. The development of hydrofoils, which had been pushed ahead during the war, was first taken up in the Eastern Bloc and later also in the USA, in order to realize the dream of an all-weather, heavily armed speedboat. The concept of hydrofoils is quite old. Nevertheless, it could not be tackled until the 1920s. Only at that time were powerful and light propulsion

*(Above) The „Arrow" („Pfeil") was a gas turbine boat built by Vosper in England. This one-off could reach up to 50 knots, making it one of the „hot rods" in the Navy's arsenal. However, the enormous fuel consumption and extreme speed, which was not tactically necessary, led to the sale of this boat to Greece.*

engines available. If you want to leave the water with the hull of your boat to reach higher speeds, you must first learn to save weight. Heavy propulsion engines are absolutely useless for this. Initially, gasoline engines were used, which originated in aircraft construction. But gasoline is too flammable a substance

*The USS "Detroit (LCS 7) of the "Freedom" Class is running at a speed of above 40 knots. This kind of warships have to be designed carefully, because any miscalculation should cause a siginificant loss of performance. So the design of this kind of ships must be supported by very sophisticated CFD software and many model tank tests.*

*(Right) John F. Kennedy was commander of the speedboat PT-109 (pictured right), which was sunk during fighting in the Guadalcanal area on August 2, 1943. Kennedy and the survivors of his crew were recovered 6 days later.*
*Researcher Robert Ballard searched for and found the wreck of PT-109 in 2002. The later president can be recognized by his typical hair bun on the right side of the picture.*

for the sailor to make friends with. Zeppelin construction, which in Germany in the period after the First World War led to technological peak performance, was not a success. World War I, had also given birth to light and powerful diesel engines. These were the ideal propulsion for the fast boats of the Kriegsmarine. Diesel oil ignites less easily than gasoline, the engines operate more economically and are also less sensitive - a propulsion solution had finally been found that has been consistently developed to this day.

At some point, the development of planing hulls was exhausted. Only small boats up to about 15 meters in length could reach extreme speeds of up to 60 knots. But these vessels were neither seaworthy nor capable of carrying the heavy equipment and armament now required for a speedboat. Other concepts had to be found. In addition to the planing hull, the hydrofoil was the most successful. Its technology is based on the idea of lifting a hull completely out of the water to reduce its resistance.

This is accomplished by airplane-like wings located in the water. Another concept has been the use of air as a „separating agent" from the water: here there are several different concepts, namely the amphibious hovercraft, the SES (Surface Effect Ship), which combines the catamaran hull with an air cushion stored in between, the air-lubricated ship (Air Lubricated) and the ground effect vehicle. Hydrofoil boats and air cushion boats have become so important, that they have each received their own chapters.

If you want to go fast, you should have powerful engines. This truism applies equally to land vehicles and to ships. Probably the most powerful ship propulsion system built to date is that of the US aircraft carrier „USS Enterprise". It is said to be able to achieve a power of about 360,000 hp through eight nuclear reactors - enough to accelerate the USS Enterprise to more than 45 knots. Officially, the U.S. Navy admits to only 280,000 horsepower. Probably, the extreme power output put so much

*(Below) The American gasoline-powered Packard speedboat engine was a licensed version of the British Rolls-Royce Merlin Mk II, the engine of the „Spitfire. V-engine with a total displacement of 40 liters, it produced about 1,500 hp at 2,400 rpm. It could rev at a maximum of 3,000 rpm. Sufficient air to breathe was provided by 48 valves and a shaft-driven supercharger.*

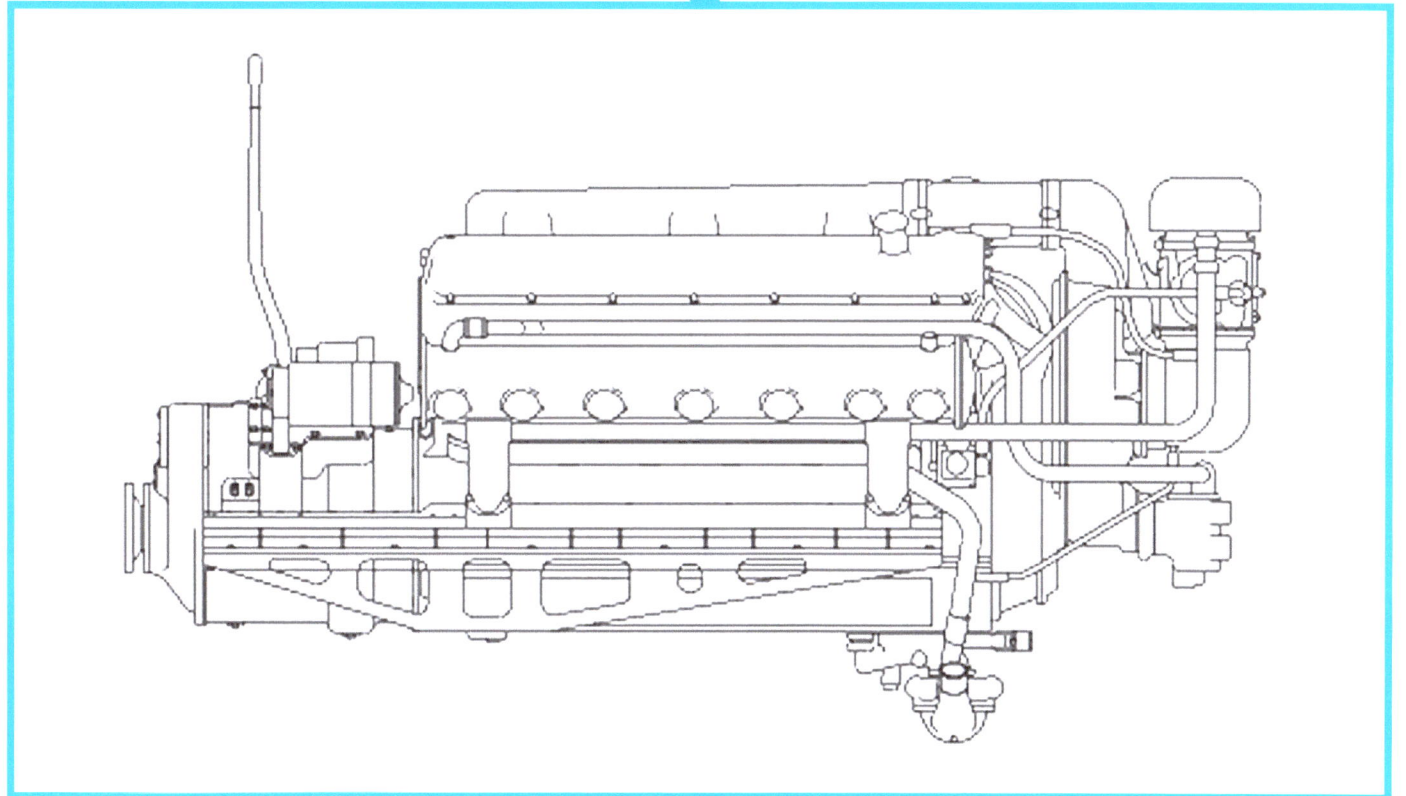

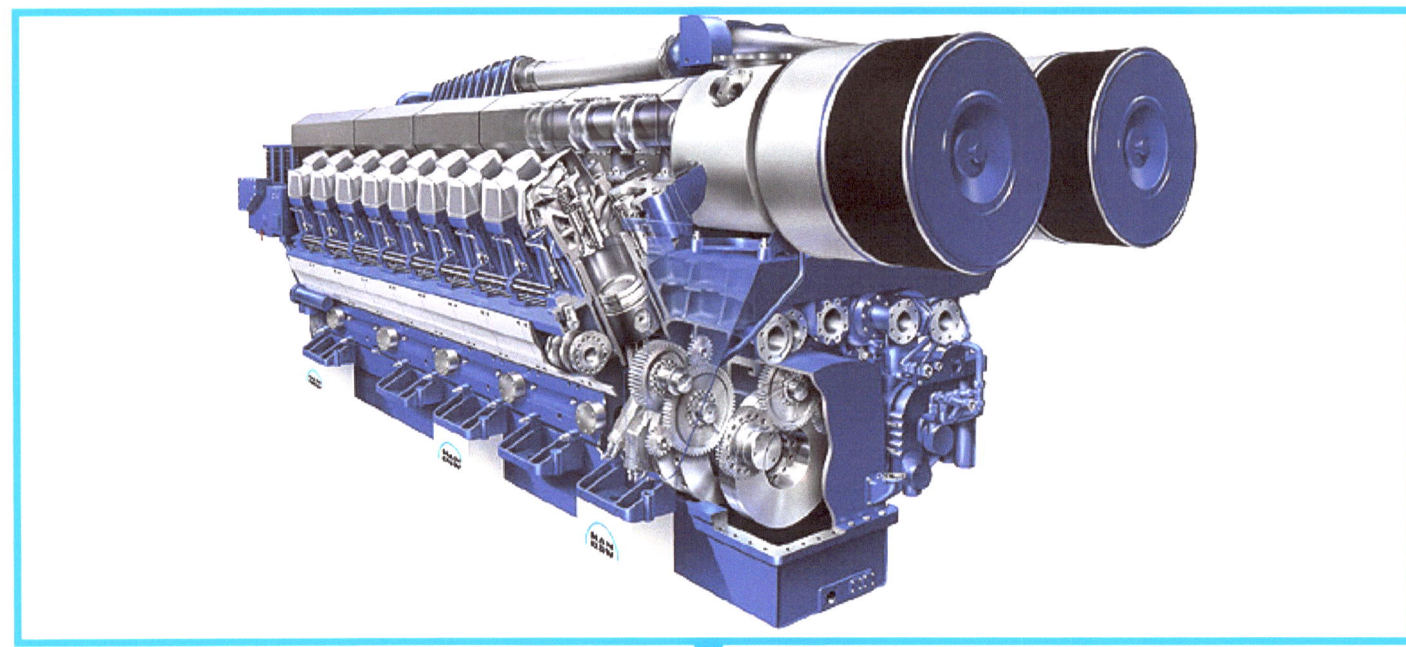

(Below) This MAK/Ruston RK 280 engine is one of four onboard of an INCAT high speed catamaran - for example the "Spearhead" as shown later in this book. The motor can produce a power output of about 9,000 hp.

strain on the propulsion system that it was throttled back. There is also no escort ship in the U.S. Navy that could follow a 45-knot aircraft carrier.

The later-built Nimitz-class carriers have only four reactors and put out 280,000 hp. The British motor torpedo boats (MTBs) were each equipped with three gasoline engines ranging initially from a total of 3,300 hp and later up to 4,200 hp. The manufacturers of these engines were initially Isotta-Fraschini and later Packard or Rolls-Royce. Such an engine had 12 cylinders with a displacement of 57 to over 60 liters. Later, compressors or turbochargers were fitted. Hardly silenced, these drives developed an incredible noise that German speedboat crews often heard from afar.

The German fast boats were initially equipped with MAN engines of 7 cylinders each, but later mostly with Daimler-Benz engines of 2,000 hp with 20 cylinders each. These engines proved to be very successful and reliable. The engines of the series MB-501 engines laid the foundation for today's market-dominating high-speed diesels from MTU, which are installed in fast boats and fast ferries all over the world. The German boats were larger and heavier than the British ones, and with three engines they needed more power. The diesel engines had a lower RPM than the gasoline engines and therefore produced a more whirring noise that was less audible.

Today, the use of gas turbines such as the Rolls-Royce „Tyne", the „Proteus" or the Alision LM 2500 on military ships is generally common.

In later times, hydrogen and electrical energy will probably take a dominant role as fuels. Very bold visions even envisage nuclear fusion as a future energy source for ships. But this can most likely only be assumed for the last quarter of this century at the earliest.

(Above) A view of a German MB-501 diesel engine. This was much larger than the Allied gasoline engines, but also made more power. Up to 2,000 horsepower was generated by 20 cylinders. But its greatest advantage was the poorer flammability of the diesel oil. This meant that the engine crew still had a certain chance of survival in the event of a hit. Their British counterparts were much worse off. There are many stories about terrible flame infernos.

*A high-power pack: Here, a MTU 4000 diesel engine is shown. Today, the state-of-the-art diesel powers fast ferries and other high-speed vessels, developing power of 1,920 to 4,300 kW (2,574 - 5,766 hp) with cylinder counts from 12 to 20. The engine is often cooled with destilled water via heat exchangers.*

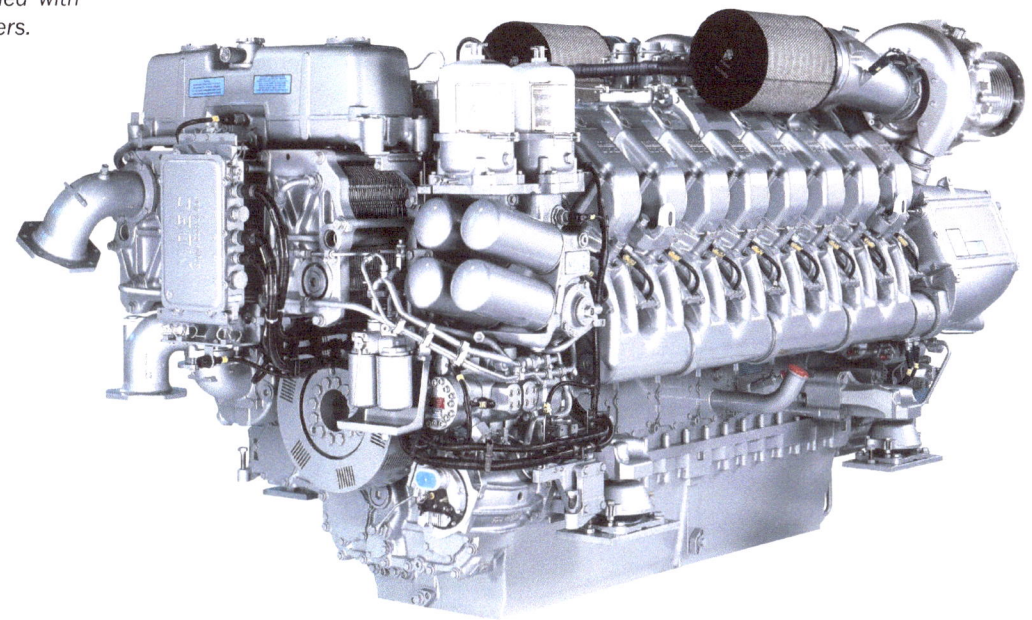

*The Rolls-Royce Marine Trent seen here is currently the most powerful gas turbine unit for ships on the market. It can develop up to 36 megawatts - equivalent to 48,276 hp. Despite this, the soundproofed unit shown weighs only about 20 tons.*
*The turbine unit is palletized for a quick remove and installation via large patches in the decks above.*
*It only take days to change a unit. modern warships are too expansive to spend a huge fraction of their lifetime in a shipyard.*

(Above) An impeller (tunneled propeller) of a large water jet unit.

(Right) The engine room of a modern large catamaran - one of the engine rooms in each hull.

# When fast boats learn to fly

*(Above) An American PHM (Patrol Hydrofoil Missle) of the Pegasus class races through the sea at about 45 knots, unimpressed by the swell. This most sophisticated type of hydrofoil to date has set the standard for performance and technology for decades to come.*

*However, its history has proven that not everything that is technically feasible is also practical. High procurement and operating costs brought this ground breaking weapon system to a premature end.*

The history of hydrofoils records a great many experiments and test runs until it was possible to arrive at efficient and reliable hydrofoil designs. It is not enough to simply use an airplane wing as a model and attach it under a boat hull. In fact, this type of boat operates on the boundary between two elements, air and water, both of which influence the physical processes at the wings and their supports. In addition, there is the sea state that the boat also has to cope with.

But the effort was rewarded. In 1942, the largest hydrofoil for a long time was launched and successfully tested on behalf of the Kriegsmarine. It was named „VS-08" and also „Schell 1." It was designed to transport a light tank or supplies quickly to a coast, but armament had also been planned. Further developments foresaw attack boats, which in combination with Hitler's small submarines should defend the coast of Europe against the expected invasion of the Allies. The invasion came in 1944, but none of the futuristic futuristic wonder weapons were ready to stop them - another failure.

For understandable reasons, the development of hydrofoils in Germany in 1944 and 1945 quickly petered out, and work began again after the war in Switzerland under von Schertel's leadership with the founding of „Supramar AG". Since then, however, Supramar has developed almost exclusively civilian vehicles. With great success, the foundation was thus laid for a new category of merchant shipping - fast ferry shipping. Supramar has since participated in a number of tenders for the development of military hydrofoils for NATO countries, but has not achieved any results.

On the other hand, the U.S. Navy was very fond of military usability and financed a lot of trials, which also led to successes. The construction of six „PHM" fast patrol boats - also known as the „Pegasus" class - marked the technical peak in this field for the time being. A PHM is an aluminum-built vessel about 45 meters long that rides on two fully submerged hydrofoils. These are designed to be retractable to enable the vessel to call at shallow harbors.

The PHM, which was developed and produced by Boeing Marine Systems, can reach up to 50 knots. However, there are narratives from former crew members that cite a figure of 60 knots. The PHMs were powered by a General Electric LM-2500 gas turbine with about 18,000 shaft horsepower. They were still able to maintain a speed of 40 knots in wave heights of over

*(Above) Research into new technologies sometimes involves trying out the unusual. This old DUKW amphibious truck was given „speed" with the help of a gas turbine and three retractable fully submerged wings. The experiment worked, but it was probably not the right way to create a new vehicle for landing troops. The „Flying DUKW" was not pursued as a project after the tests („DUKW" means „Detroit United Kayser Works = manufacturer of this vehicle).*

3.5 meters without it becoming unbearable for the crew. This performance, which could not be achieved by any conventional speedboat, was made possible by an autopilot system that flew this ship through the waves fully automatically. The helmsman only had to set the course and adjust the „flight altitude".

When the PHM was to enter a sharper turn, the autopilot automatically initiated some heeling like an airplane to keep the lateral acceleration in the turn under control in the turn. There was also an emergency stop program that allowed a PHM to „land" and brake down as quickly as possible.

To this day, the PHM remains the most technically advanced and powerful missile speedboat built to date. It was far superior to all other designs of the 1970s and 1980s. Equipped with up to eight Harpoon guided missile launchers and a 76mm gun from Oto Melara, it was also able to it could also strike hard. Despite its performance superiority, it failed to really convince either the U.S. Navy or its other NATO partners.

*(Below) Enrico Forlanini's first experimental boat from 1910. This is how it all began.*

Since the PHM had been developed by an aircraft manufacturer, it was also equipped accordingly with lots of expensive and complex aviation hardware. In addition, development had taken a long time and cost a lot of money. This now had to be recouped. The PHM did not really fit the mission concepts of the US Navy. It would not really have been able to accompany the large aircraft carrier groups operating globally, because its range of about 600 nautical miles was too short. Only a conflict close to shore and in restricted waters would have been the right environment for the PHM. But it is precisely this environment that the carrier groups and the nuclear submarines of the U.S. Navy avoid like nothing else, because there they would lose every advantage - namely, being able to hide in the vastness of the ocean.

The PHM was offered by the U.S. to NATO allies. Boeing had high hopes for good business in Europe, but the German Navy and others withdrew from the program. Hydrofoils have the decided disadvantage that they can only go very fast or very slow. Medium speeds are not feasible.

The wings of a PHM do not begin to develop enough lift to take off until 25 knots. Before that, they generate mostly drag. Auxiliary propulsion systems allow it to reach only about 11 knots - just enough for port manoeuvres. Thus, the PHM lacked the flexibility of conventional missile speedboats, which were slower but also handled well. The PHM was very expensive. The German Navy could have purchased only a handful. The six PHMs were stationed by the Navy in Florida. This allowed them to threaten a little in the direction of Cuba. Their hour came when the U.S. government declared a „war on drugs."

The U.S. Navy was called in to help federal authorities hunt down drug smugglers, who were especially active in Florida in the 1980s. The smugglers' preferred mode of transportation was offshore racing boats. These boats were powered by American „big-block" engines to speeds of nearly 60 knots. They were used to break through Coast Guard chains of control and take over drugs on the high seas. The conventional Coast Guard boats could no longer keep up. When PHMs became involved in this competition, the risk to smugglers increased. Still, it re-

(Right) This foil of a Russian type boat runs very fast through the water, so it must be manufactured of high tensile steel.
Most foils can cut small pieces of wood. But plastic panes and other weak materials are a natural enemy of any hydrofoil vessel.
Il a foil bumps against very hard and heavy debris it will normally break at planned break points to prevent very serious damages on boats hulls structure. Such incidents have happened from time to time. Amazingly, the biofilm of algae can survive the high speed through the water.

## The propulsion system of the PHM:

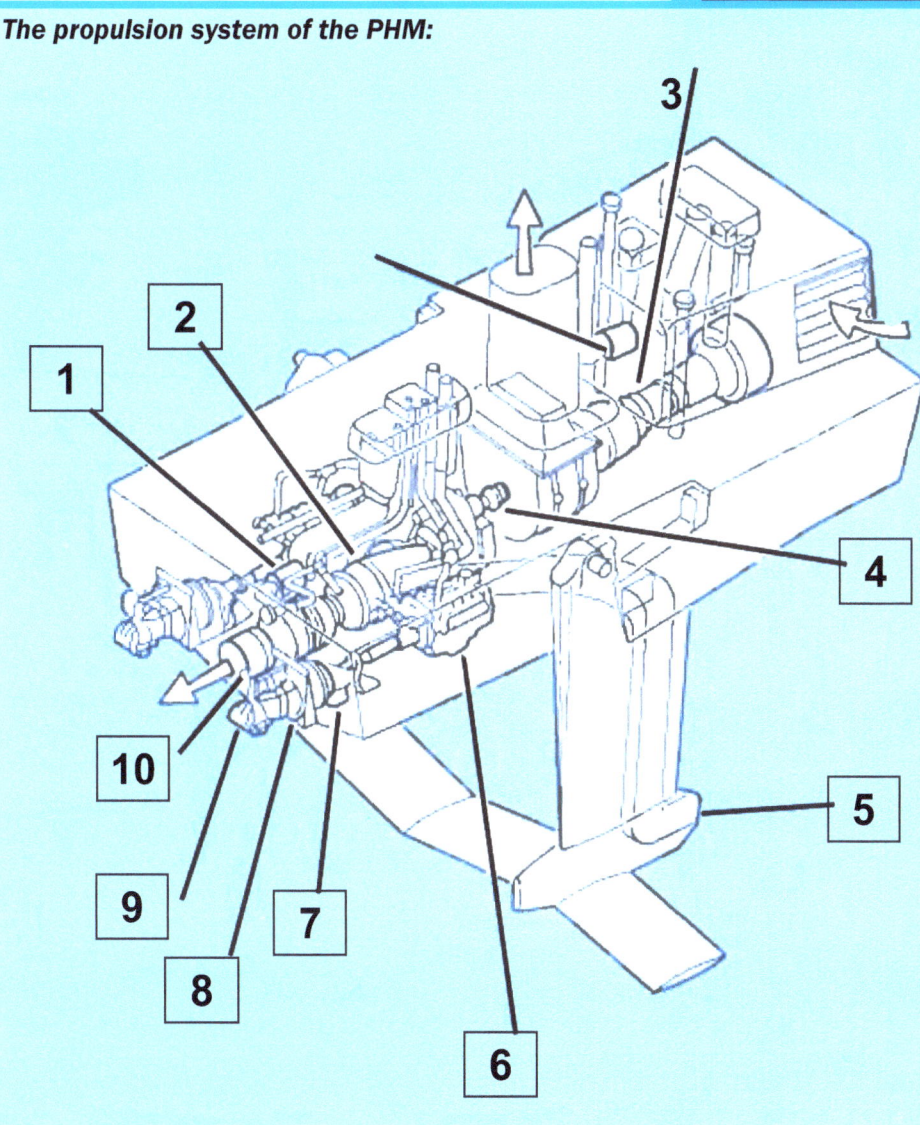

The gas turbine drives a water pump that draws in water through an opening in the wing struts and ejects it rearward through the main jet nozzle. The recoil acts as the driving force. For slow travel when the wings have been retracted, there is an auxiliary propulsion system consisting of two diesel engines and associated smaller jet engines. These only work when the PHM is not „flying".

| | |
|---|---|
| 1 | Control unit |
| 2 | Main drive (jet pump) |
| 3 | LM-2500 gas turbine |
| 4 | Main shaft with gearbox |
| 5 | Rear wing support and water intake opening |
| 6 | Auxiliary drive (diesel engine) |
| 7 | Auxiliary drive intake opening |
| 8 | Auxiliary engine water jet pump |
| 9 | Jet nozzles with control flaps (auxiliary drive) |
| 10 | Main jet nozzle |

mained only an interlude. The PHM boats were mothballed in the early 1990s and sold for scrap early in the new century. Only one has survived to this day, the USS 'Aries' - now a museum ship.

This technology, which seemed so promising in the beginning, ended up being very disappointing due to high costs and its vulnerability.

*(Above) The autopilot of this PHM automatically banks the ship to avoid harmful G-forces. The speed is very high for a boat at about 50 knots. It behaves more like an airplane than a ship.*

*(Below) The anatomy of a PHM shows a powerful gas turbine engine, weapon systems, sensors and control systems and nethertheless the small crew quarters.*

*(Above) A Number of bad ideas, but still very racy. The Carl XCH-4 reached up to 78 knots (144 km/h) in 1953 as a world record speed on behalf of the US Navy. The problems of he water propeller propulsion with the cavitation to be expected were simply circumvented by air propellers.*

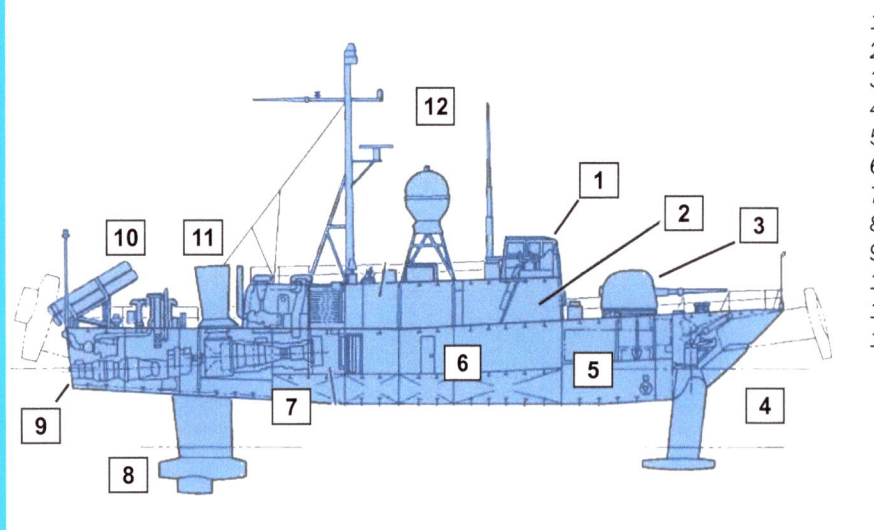

| | |
|---|---|
| 1 | Bridge |
| 2 | Combat Intelligence Center (CIC) |
| 3 | 76mm gun |
| 4 | Front wings for retraction |
| 5 | Ammunition storage 76mm |
| 6 | Living quarters |
| 7 | Gas turbine (main propulsion) |
| 8 | Rear wings with water intake |
| 9 | Auxiliary engine |
| 10 | Harpoon launch container |
| 11 | Gas turbine exhaust tower |
| 12 | Air and sea target radar |

# Overview of other some military hydrofoils

In the conservative world of the military, hydrofoils have had very limited success. The main obstacles to their use are high construction costs, low operational ranges and the high sensitivity of the extremely lightly built craft.

### Sparviero (Italy)

| | |
|---|---|
| Year of construction: | 1974 |
| Length/width: | 24,6 / 12 m |
| Displacement: | 62,5 ts |
| Power: | 1 X gas turbine 4,500 hp |
| Speed: | Max. 50 knots |
| Armament: | 1 X 76 mm cannon |
| | 2 X OTOMAT FK. |
| Status today: | Out of service |

### Shimrit (Israel - license Grumman USA)

| | |
|---|---|
| Year of construction: | 1982 |
| Length/width: | 25,6 / 7,3 m |
| Displacement: | 105 ts |
| Power: | 2 X gas turbine 5,364 hp |
| Speed: | Max. 48 knots |
| Armament: | 1 X 30 mm MK |
| | 4 X Gabriel or Harpoon FK |
| Status today: | Out of service |

### HMS Speedy (United Kingdom)

| | |
|---|---|
| Year of construction: | 1980 |
| Length/width: | 27,4 / 9,5 m |
| Displacement: | 117 ts |
| Power: | 2 X gas turbine 3,780 hp |
| Speed: | Max. 43 knots |
| Armament: | none |
| Status today: | sold in 1986 and converted to civilian ferry. (Today in service in Hong Kong.) |

### Antares (Soviet Union / Russia)

| | |
|---|---|
| Year of construction: | 1983-1989 |
| Length/width: | 40.3 / 8 m |
| Displacement: | 220 ts |
| Power: | 2 X gas turbines 3,780 hp |
| Speed: | Max. 65 knots |
| Armament: | 1 X 76 mm gun and smaller weapons. |
| Status today: | Out of service |

# Small warriors

*(Above) These so-called „Semi-RIB" of the U.S. Armed Forces are typical representatives of modern small combat boats. „Semi-RIB" means „Semi rigid hull inflatable boat". A fixed hull made of fibre reinforced resin or aluminum is fitted with an flexible tube around the outer rim, which serves as a capsize protection and fender. It can often be made smaller by deflating it to make the boat more easily transportable. These types of boats come with and without cabins and with inboard or outboard engines. They are usually very fast and inexpensive. Their seaworthiness is very high.*

*For small combat units, these boats are the ideal means of transportation because they are transportable by cargo aircraft such as the Lockheed C-130 Hercules and can be jettisoned on parachutes. Armament often includes heavy M1 machine guns, M-60 and often anti-tank missiles. Sometimes an M-79 machine grenade launcher is also used. The SEALs can be deployed worldwide very quickly with such boats through airlift and parachute drop without problems. In the times of covert conflicts, such units are the right „weapons" against terrorists and criminals.*

In the secret world of special forces and intelligence, fast and light sea craft have always played an important role. A typical clandestine task for World War II torpedo boats was dropping off agents and saboteurs on enemy shores. The small size protected against premature detection and the speed facilitated escape. Only submarines were even better suited for this work, but these required deep water to take advantage of their diving capability. The Royal Navy records in its chronicles numerous missions during World War II in which young lieutenants were assigned as speedboat commanders to drop off or pick up silent, hardened men and women on France's canal coasts.

They were inculcated never to breathe a word about it, and that was it. A brief touch of adventure as a change from routine duty? It certainly wasn't like that, but it made a good subject for the novels of numerous naval writers later on. In those early days, people were forced to improvise. The boats and equipment were inadequate, their crews were young people with little experience, and the secret agents themselves were not born as such.

In modern times, almost everything has been professionalized and even the equipment has been developed and adapted quite specifically for this purpose. A good example are the Royal Marine Commandos and the US Navy SEALs (Sea/Air/Land), who are trained extremely hard for years as professional soldiers for these tasks.

Only the fittest, toughest and most talented soldiers are selected from numerous applicants. During months of training, they undergo unimaginable hardships without being able to afford a single mistake. This means exclusion from the training. Accordingly, these men and women are held in high esteem by the regular troops.

Special soldiers also require special equipment. Characteristic of these units, often called „snake eaters", is that they are fully utilized even in peacetime. Covert reconnaissance missions, crime fighting, object and personal protection or even hostage rescue are fields of activity that are commonplace for these people. In addition, there is a certain percentage of „black" non-legalized missions, which are never discussed or recorded. These

*(Above) A Dockstavarvet CB 90 of the Norwegian Navy lays into the curve. The very sophisticated hull shape of this type of boat makes for excellent handling. In front of the bow, the longitudinally divided bow hatch can be seen, and below it the drawbridge-like ramp by which landing parties leave the boat. Since the draft of the nearly 18-ton assault boat is very shallow, it can land troops on almost any type of coast.*

missions do not only originate from the imagination of thriller writers, but actually take place.

All special forces love secrecy and appearing on the scene as inconspicuously as possible. Therefore, quiet means of transport are preferred for transportation. In many operations of maritime units, support by submarines is resorted to. But sometimes things have to move faster. Then one is dropped on the spot by a light boat on a parachute and can do his job.

As armament, the soldiers' personal weapons such as assault rifles or submachine guns are usually sufficient. If more firepower is needed, one helps oneself in the „warehouse" of the land soldiers: Light rocket launchers, anti-tank missiles, light and heavy MGs or a machine grenade launcher. Often little is enough to achieve great effect. After all, nobody wants use a sledge-hammer to crack a nut

The field of these specialists is limited warfare, which means that an effect must be achieved with as little force as possible. The use of heavy equipment is frowned upon. A SEAL commando is often only noticed by a normal person when it appears directly in front of him. But by then it is already too late for enemy soldiers.

Typical light boats are open fast inflatable boats or rigid hull boats made of plastic, usually powered by powerful outboard motors. The main thing is that the boats are light and can be transported quickly. Special mufflers often ensure silent driving. Everything is painted in low-reflection camouflage colors.

But there are also special larger boats built for small landing operations. One such is the „Combat Boat 90" built by the Swedish shipyard Dockstavarvet AB. The boat was specially tailored to the Swedish coastal landscape with its countless small islets and rocky bays.

The 15.90-meter-long and 3.85-meter-wide boat is capable of carrying and launching up to 21 infantrymen at speeds of up to 45 knots with a crew of three. It is something of a counterpart to the land-based armed personal carrier (APC), as it is usually armed with three 12.7-mm caliber MGs and a grenade launcher. In addition, „Hellfire" guided missiles can be fired, which can be used against both tanks and other water vehicles. The gun emplacements of the MG can be remotely controlled from inside the boat. The ‚CB 90' is even armored against light weapons and can thus operate in a very hostile environment and very close to shore.

Anyone who has seen the manufacturer's spectacular promotional videos knows how manoeuvrable these boats are. Very tight turns and stopping from full speed in little more than two boat lengths is no problem.

For the Swedes, the CB 90 is a good way to deliver smaller infantry units very quickly and under fire to the offshore islands and headlands of the confusing archipelago. In doing so, it is less vulnerable than a helicopter unit and also costs much less.

The performance of this boat did not remain hidden for long, so that other countries such as Norway and Brazil also procured

*(Above) U.S. Navy SEALs at their normal environment. The SEAL on the right is a so-called „painter": he marks a target for a laser-guided bomb with a laser marking device. Accordingly, the bomb is already dropped a few kilometers away by a fighter aircraft. In this way, a small team with outside help can deliver a powerful and precise strike against an enemy that is weighing itself securely. The team is usually dropped from small boats, submersibles or by glider parachute on the coast.*

these boats and use them with great satisfaction. Even the German police borrowed a CB 90 when the G8 summit was held on the German Baltic coast in the summer of 2006. The CB 90 is a good example of the capability and combat power that even small boats can offer today. Some also serve as sea ambulances or pilot boats.

Dictators are often „technology freaks." This was already the case with Hitler and has continued with Kim Jong-il, and his son Kim Jong-un of North Korea. North Korea is not only known for the production of missile weapons that are undesirable in the West, but it also has a large number of infiltration, sabotage and other special forces that are supposed to infiltrate, spy on and sabotage the state of South Korea in preparation for an invasion.

Given the fact that the military budget of the country is so high that the population is partially starving, it is not surprising that these secret forces are equipped with any „gadgets" that are possible. Thus one has there small submarines, numerous high-bred motorboats and also over particularly „hot rods", like so-called „Very Slender Vessels" (VSV), thus boats, which are particularly slim. VSVs were actually invented in Great Britain and are used there militarily by the Special Boat Service, a special unit of the Royal Navy. The SBS can use these boats to carry about 4 - 6 soldiers through rough seas at speeds of up to 50 knots. The boats can even be dropped fully manned from a transport aircraft on parachutes.

VSVs represent the pinnacle of the art of building small fighting boats, but their use requires great physical stamina, A VSV usually no longer jumps over the waves, but pierces them as much as possible. Nevertheless, the acceleration values inside are considerable and it is very exhausting to sail with them.

Slim boats roll heavily and the VSV is no exception. So one should not be prone to seasickness. But special forces operatives are used to enduring uncomfortable environments. In North Korea, the VSVs have been replicated according to available evidence. As an aside - the Colombian drug kings have also discovered the VSV concept for themselves. Thus, one has already fallen into the hands of the police authorities and has been investigated. VSVs are easily built from plastic in improvised hidden yards. So a good way to cheat the ever-vigilant coast guards.

*(Right) A CB 90 enters the docking area of the British landing ship „HMS Bulkwark". It can form a base for a whole group of these boats.*

*((Below) The CB-90 can stop very abrupt using the the full power of two heavy Scania Diesels. The water jets are very effectve at reverse thrust.*

*(Below) The CB 90, despite having a shallow draft, is very spacious. Due to its enormous cruising power, its armament and the armor protection, this ship and its variants still has a great future ahead of it. It is conceivable that such small craft will soon be equipped with more powerful guided missiles, which could be dangerous even to much larger ships. Then it would be part of a very large threat to large fleets from small navies.*

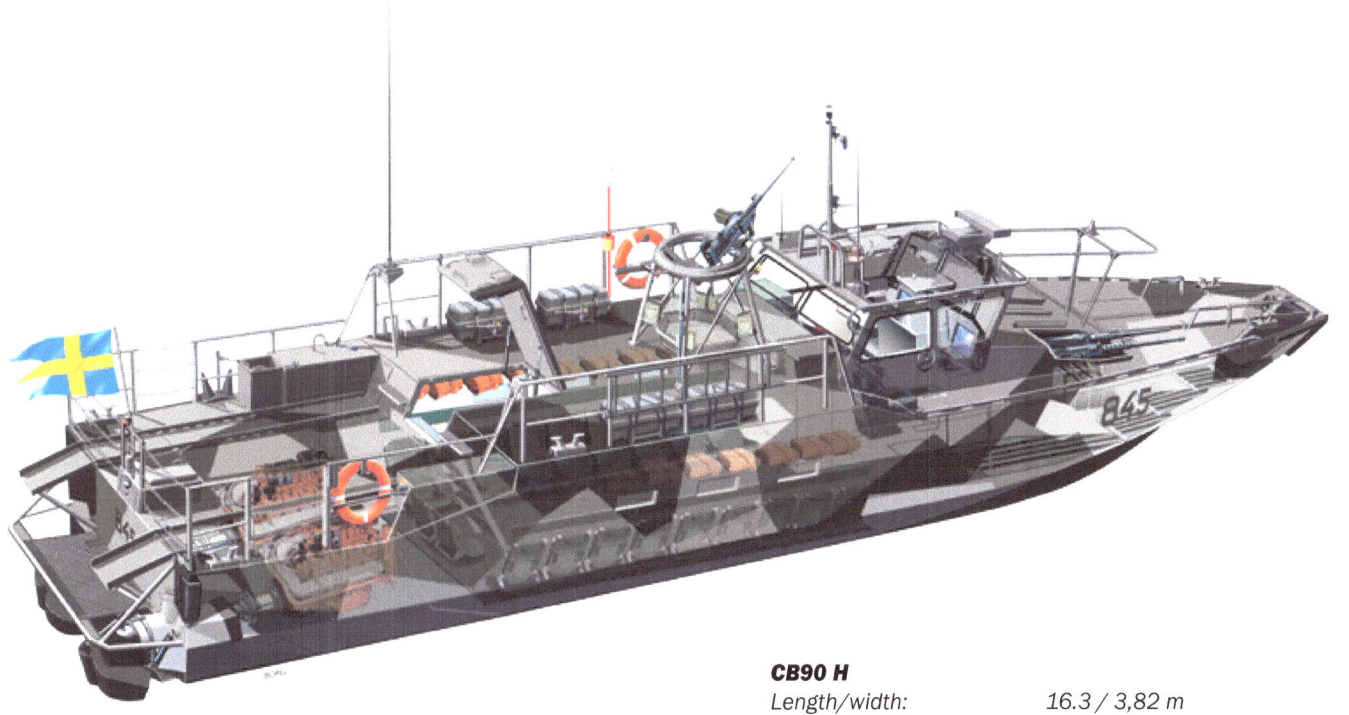

### CB90 H

| | |
|---|---|
| Length/width: | 16.3 / 3,82 m |
| Displacement: | 18 tons |
| Power: | 2 x Turbo diesels each 600 kW |
| Speed: | Max. 45 knots |
| Armament: | Machine guns, Hellfire missiles, 40mm grenade launchers and other weapons |
| Crew: | 2 - 3 crew |
| | 20 troops |

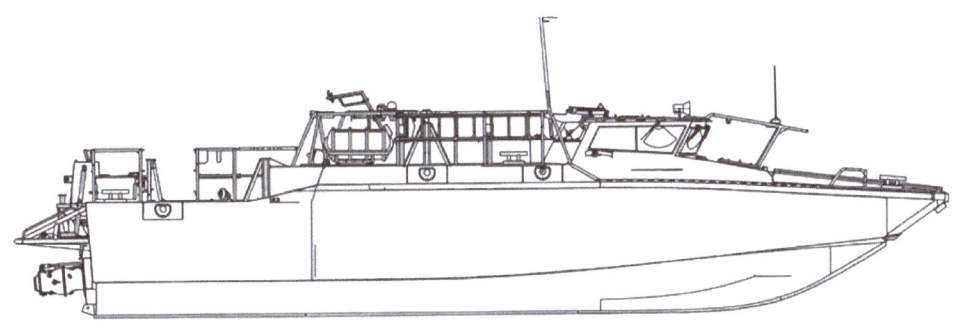
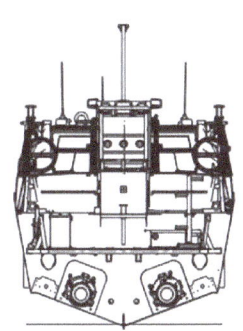
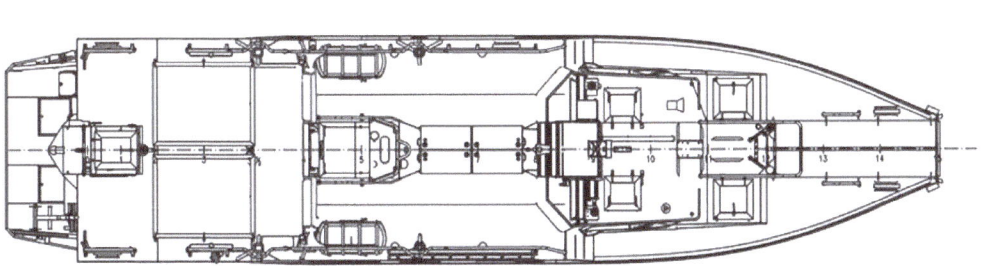
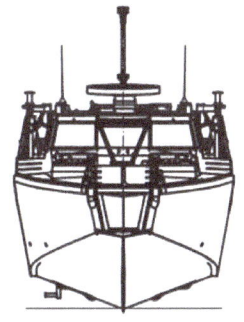

*(Below) Presumed profile of a North Korean VSV with a length of about 32 meters. It is thus already too large to be able to hide effectively from RADAR. This problem is compensated by a very high speed even in rough seas.*

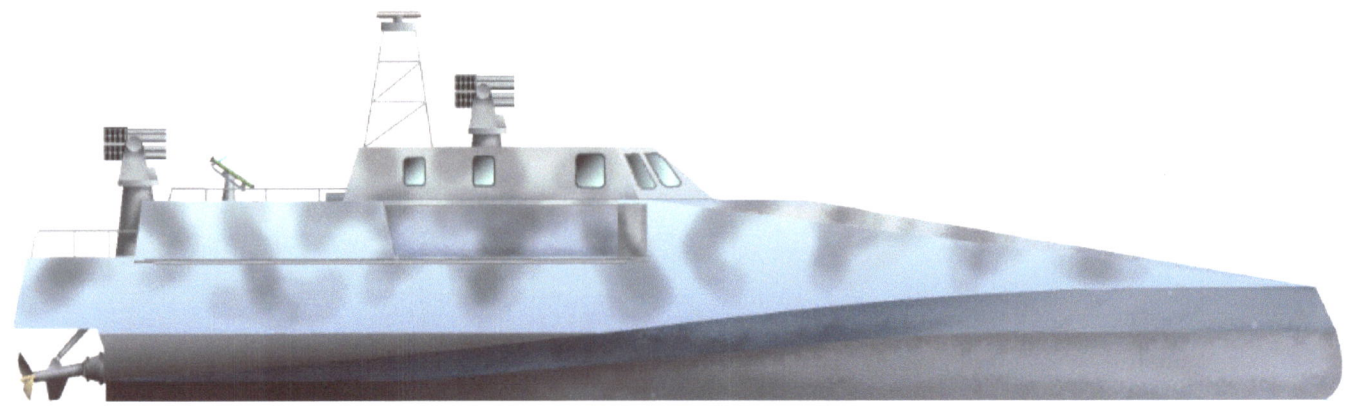

*(Below) The American „Sealion" is a U.S. Navy high-speed boat for the Navy SEALs. It is 24 m long, 4.4 m wide and carries seven people at probably about 50 knots. A jet ski serves as a tender. The boat has anti-RADAR capabilities. From a technical point of view, it is otherwise a completely normal glider.*

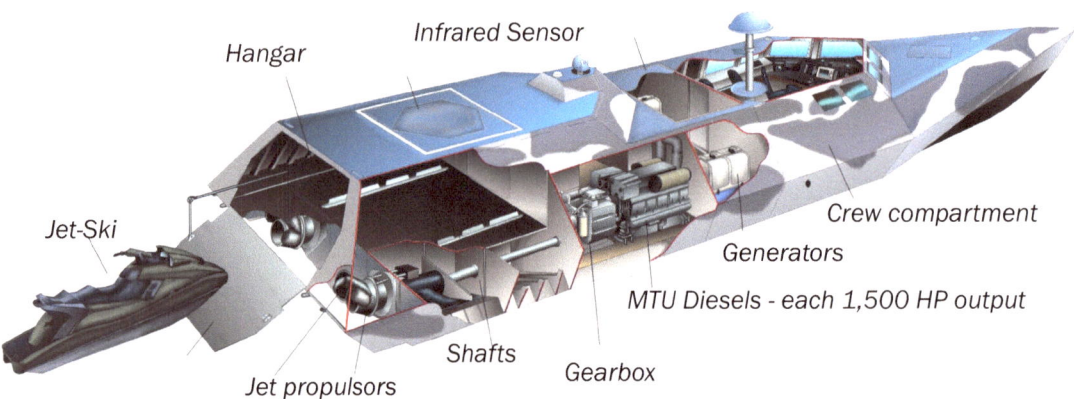

A drug-VSV seized by police forces. The propulsion is provided by two outboard motors. The hull is made of hand-laid GRP. The construction is improvised, the finish poor. It was designed for only one or two missions. It can carry many millions of dollars worth of drugs.

Today this kind of craft is seldom used by the drug lords. Semi-submersibles and submersibals are more often used due to their higher capacity and range. Also a VSV cannot hide itself as effective as semi-sub on surface against infrared and radar sensors.

The range of those small subs is often enormous. One have been caught at the Spanish atlantic coast in 2023.

# The fastest tank on the water

***A tank in this book?*** *The U.S. Expeditionary Fighting Vehicle („EFV") is almost certainly the only tank in the world that can reach 20 knots on water. It would have been a tremendous advance for the Marine Corps had it not been cancelled in 2011 because of the U.S. fiscal situation. This was a bitter hour for the Marines. The well-armed floating tank can also boast considerable performance on land. It is powered by an MTU 883 diesel that produces 2,300 hp in the water and a throttled 830 hp on land. It transforms from a tank into a speedboat by retracting the tracked gear and deploying a nose shield.*
*The Marines have searched for decades for an armored vehicle that could travel equally fast on land and water to more safely drop landing troops. A wide variety of ideas have been up for discussion over the years. Everything from hovercrafts to hydrofoils were considered. But this surprisingly simple solution won out in the end. The variable power of the engine was a key factor.*

# Riding on air

The invention of the hovercraft has been an often misjudged technological advance. For many, the spectacular, fast and impressive hovercraft was the solution to all transportation problems, and for these people it was only a matter of time before conventional ships would be replaced by the new „hovercrafts".

Other groups were more skeptical about the new technology. One would be able to use this technology militarily and civilly at best for very specialized tasks, it would be too expensive anyway, and so on. In the end, both groups were only partially right. The hovercraft was able to carve out its place in the world of armor where no one else could demonstrate similar capabilities, such as landing troops. Its high cruising speed, ability to capture beaches unchallenged by water depth or ground conditions, and high transport capacity have made it a unique offensive asset.

The history of the United States Marine Corps (USMC) is most impressive for the large number of landing operations it conducted during the World War II era. Without the USMC troops, who specialized in this type of warfare USMC troops, action against Japan would hardly have been possible. The so-called „island jumping" ultimately served to obtain a position from which heavy air strikes could be dealt out against the Japanese mainland. This was to break up the industry that kept the Japanese war machine running.

The Marine Corps had not been born with the capabilities to land troops, but they were much better prepared to do so than the U.S. Army was for the Normandy landings in 1944. The USMC military manual contained tactical instructions for forming and executing amphibious operations even before the war.

But the slow and barely seaworthy landing craft of the war no longer had a place in the increasingly fast-paced world of the postwar era. They were too vulnerable and also barely able to visit any beach at an arbitrarily chosen time. A landing with these ancient vehicles had to be carefully coordinated with weather conditions and tidal patterns. Those who have followed the history of „D-Day", the Allied landing in Normandy in 1944, know that „Ike" Eisenhower in view of bad weather reports, had only a few temporal alternatives to choose from for June 6, 1944. The ideal tide levels needed would not have reappeared until weeks later.

*(Below) The U.S. Navy's LCAC hovercraft has become one of the primary means of transportation for landing operations by U.S. forces.*

*Its speed and maneuverability make it absolutely unrivaled by conventional landing craft. (LCAC = „Landing craft air Cushion")*

*(Above) A U.S. Marine Corps M-1 tank rolls onto the beach from the LCAC. Such a tank can greatly facilitate the work of a landing force. It has high firepower, mobility, and strong armor, allowing it to break through almost any defensive line on the beach. Then, together with other tanks of the same type, it can advance very quickly (about 60 km/h) into the hinterland.*

So the USMC had been dreaming since World War II of a landing craft that could move rapidly from the transport ship (LSD or „Landing Ship Dock") through a choppy sea to a beach. It was to be able to carry heavy battle tanks in the process so that it could land the right „punch" on the enemy from the start. The Corps never forgot how often its soldiers had to lie on the beach as if pinned down for hours in machine gun fire from the defenders, until only in a later phase of the landing could the tanks roll onto the beach.

Another problem was the danger posed to the valuable transport ships by the defenders' defensive fire. In the campaign against Japan, American battleships, cruisers, and destroyers often hammered Japanese bunkers for days before landing troops to pulverize the defenses.

But the Japanese quickly learned how to conceal and secure their own guns. Thus, landing troops were then often given an unpleasant surprise by undestroyed Japanese defenses. If they were heavily enough armed, they also sometimes took the landing ships and transport fleets under fire.

Today there are no battleships armed with equally heavy artillery. The modern guns of frigates and destroyers today are at best 13 cm caliber, which is really just „pop guns" compared to the 40.6 cm guns of an Iowa-class battleship. Today's landing dock ships, on the other hand, are much larger and more expensive than their wartime counterparts ever were. Their loss would be catastrophic. Modern landing forces can only rely on aerial bombardment to hold down defenses. But that is simply not enough.

The Marines preferred to have to rely only on themselves. Their own main battle tanks were their first choice when it came to making a stand on the beach and in the rear. The Marines have also maintained their own air force since World War II. Today, their main weapon is the F-35 variant capable of vertical takeoff, and the USMC has its own artillery units that - once ashore - can continue to hammer away at the enemy.

The technology of the hovercraft finally made it possible to create the desired landing craft, that LCAC (Landing Craft Air Cushion). When the S.R.N. 4 canal ferries in Great Britain proved that large and powerful hover transporters were no longer science fiction, the USA and the then USSR began to work intensively on the use of this technology for the military. The landing of troops was the primary focus. Both armed forces had to deal with similar problems:

The USSR was planning a large-scale landing operation on the German and Danish Baltic coasts in the event of an East-West war. But there were the air and naval forces designated by NATO to defend those coasts. This force, composed of speedboats, submarines and low-flying fighter-bombers, had been tailored precisely to defend against ship formations approaching the coast. Slow landing craft and transport ships would have had little chance of forming a bridgehead for heavy formations in the northern German areas of Travemünde, Kiel or Flensburg. So plans were made to land assault troops by fast craft that could approach the coast taking advantage of the surprise effect.

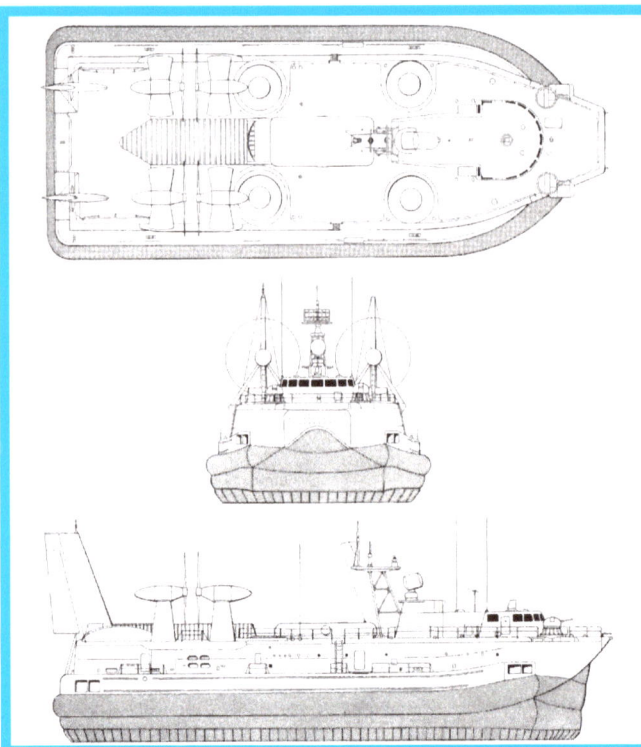

*(Above) The Soviet "Aist" hovercraft was one of the largest hovercrafts in the world at the time.*

The soldiers and tanks released by them would have settled on the shore and secured the landing head of the main army heavily armed with tanks and guns.

Soviet designers first created a hovercraft landing ship, the „Aist", and several smaller craft for this purpose. With a total mass of 298 tons, the „Aist" was one of the largest hovercrafts of its time and could carry about 80 tons of load at 50 to 60 knots. The cargo included PT-76 and BMP light tanks and up to 200 infantrymen. The „Aist" were formidable vehicles with great capability and were surpassed only by the „Zubr" class, which is still used today. Crafts of the Zubr class were even exported to Greece. A Zubr has a maximum mass of 535 tons and can carry over 130 tons of cargo. Unlike the Aist, it is also armed with offensive weapons such as artillery rocket launchers and anti-aircraft missiles. Its lift and propulsion system is powered by 5 gas turbines of about 10,000 horsepower each. The Zubr is undefeated as the largest hovercraft ever built. Its noise performance is extreme and it can reach about 50 knots unloaded. A single vehicle cost about $100 million in the 1990s.

Due to the very offensive nature and extreme procurement and operating costs of the Zubr, it is unlikely that more than the nine in existence today will be built. Despite this ship's large-scale appearance, one thing it cannot do is carry heavy battle tanks.

Its vehicle deck is only suitable for carrying relatively light vehicles such as infantry fighting vehicles or the PT-76 light amphibious tank. This light tank is only weakly armored and represents only one thing for a modern heavy battle tank like the M1A2 - an easy target.

The American LCAC is intended to serve a similar purpose to the Zubr. But it pursues an absolutely different strategy: as a replacement for the heavy landing craft type LCU (Landing Craft Utility) used so far, it is to be compatible with all mother ships currently in use for landing units. These have a dock at the stern that can be floodable dock at the stern that can serve as a safe haven for landing craft. This is where the landing boats are loaded and also transported. The LCAC is capable of carrying a USMC M1-A2 tank or two complete 155 mm howitzers with crews and towing vehicles at about 40 to 45 knots and dropping them on about 80 percent of the world's beaches.

It is unaffected by underwater obstacles, mines or impassable terrain and can even travel inland on shallow river courses to the drop-off point. In doing so, the fleet of landing mother ships remains out of sight beyond the horizon. The LCAC has greatly improved the USMC's transportation capability. Nevertheless, it is maintained and operated by the U.S. Navy.

The job of LCAC commander is reserved for a „chief" (chief petty officer). The LCAC chiefs of the U.S. Navy's two LCAC units take great pride in the fact that, as noncommissioned officers, they command a vehicle that is as expensive as an F-18 fighter, which is basically flown only by officers.

The LCAC has finally given the Marine Corps the mobility it has always wanted. Besides the Navy's 91 LCACs, only a few hovercraft are still operated in troop service around the world. All are much smaller and are mostly used to transport small units to special operations and command operations. The Royal Navy's Marines operate a small flotilla of Griffon 2000-TDS hovercraft, which serve as fast carriers for heavily armed squads of Royal Marines or Special Boat Squadron. The hovercrafts have access to areas where navigation by inflatable boats and landing craft is no longer possible. For example, they are deployed in the Arctic Circle or race through the swamplands of Borneo or Malaysia. As is well known, the Union Jack flies in many places around the world. These little hovercrafts dwarf the Zubrs or the LCACs, but the Royal Marines wouldn't want to be without them. They can even be disassembled and loaded onto cargo planes, making them globally deployable.

Hovercrafts were once considered a breakthrough technology for fast ships, but it has since been shown that comparable results can be achieved with less effort. They only really have an advantage where they can exploit their amphibious properties.

*(Left and following page and below) Impressions of the Zubr hovercraft. This monster is today a kind of „white elephant" - impressive in every respect, but too expensive to operate and procure. There must have been special reasons for the Greek government to purchase four of these vehicles.*

But the energy required to operate them and the construction costs are far too high compared to more conventional designs. The giant Russian Zubr hovercrafts and other ‚monsters' of this type will probably die out eventually, like all great dinosaurs. They will be fondly remembered as an example of how, in spectacular fashion, a type of ship was born that had completely novel characteristics and sailing performance.

*(Below) The Layout of the "Zubr" Class shows several waepon stations on the open deck. This units are the heaviest armed hovercrafts on the planet. The weapon system comprises gatlings guns against air threats and rocket launchers against land targets*

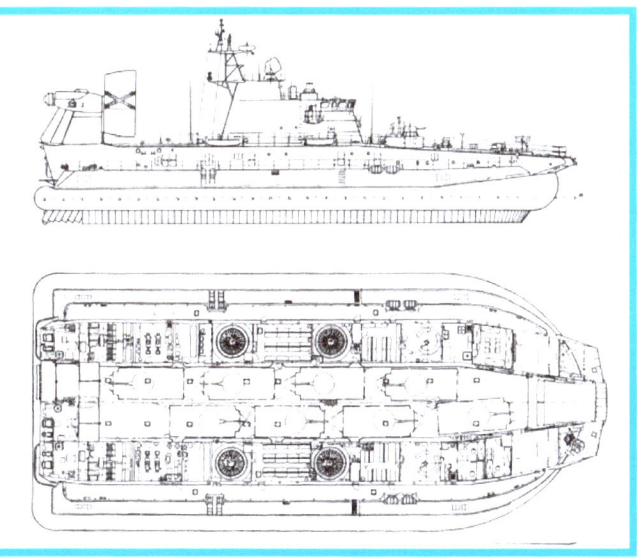

(Left) Two LCACs have set up shop on a picturesque tropical beach. The Corps operates worldwide. Often a ‚Marine' comes to places in the world where others would love to spend their vacation.

(Above) These Zubr hovercrafts observed near Kaliningrad in the Baltic Sea certainly look impressive. But even one or two fighter-bombers can sink them very quickly. It's 45 knots on the water versus about 500 knots in low-level flight.

(Below) This is a Griffon 2000-TDS, shown here patrolling the port of Basra under the Royal Marines flag. This versatile vehicle is a reliable transporter for small units.

(Above) This Swedish Griffon 8000 is already capable of carrying smaller motor vehicles. It is still powered by compact and economical diesel engines despite its size of about 19 meters.

(Below) The USS San Antonio, a state-of-the-art landing dock ship, with its cargo consisting of two LCACs, among others. The hovercrafts save a lot of time because the dock no longer has to be flooded to launch them. The general layout of the "LSD" was found in the WW II to support the lading operations in the Pacific.

# Highspeed in military logistics

*(Above and previous image page) The Australian HMAS Jervis Bay was the first fast ferry catamaran to be used for military purposes. As a troop carrier, the concept was a great success from the start and prompted thinking about more advanced applications.*

Transportation by sea has become a problem in the era of rapid reaction forces that can respond to crises within days or even hours. Conventional cargo ships are simply too slow to reach the scene of an incident over global distances of several thousand nautical miles in just a few days.

Just after the tsunami in December 2004, arguably the most momentous natural disaster in recent history, it took several days before even fairly fast aircraft carrier units could get into position off disaster areas in Thailand, Sri Lanka or Indonesia to bring help to the victims through their extensive equipment. In the early days, aircraft alone were able to provide the needed transportation for relief supplies. Around 2004, the U.S. Army had its own large high-speed ship, the „USS Spearhead." This 98-meter catamaran had been leased as a multipurpose logistics ship from the Australian shipyard Incat Tasmania Ltd.

With a top speed of 38 knots and a range of several thousand nautical miles, the aluminum twin-hulled ship was able to carry a payload of double-hulled ship could carry a payload of several hundred vehicles, military containers and several hundred personnel. She was immediately put to sea and crossed the Pacific in just a few days. The volume it carried was so large that only a large fleet of C-17 transport aircraft, each with a payload of 40-50 tons per aircraft, would have been able to compete with it in long-range flight.

The tendency to use modern civilian ferries for troop logistics was first practiced in the Falklands War. There, it was container ships and North Sea ferries, hastily requisitioned, that took over the transportation of the more than 5,000 British ground troops and their vehicles. The first fast ferry in Navy colours was the „HMAS Jervis Bay". This 86 meter long catamaran had been chartered from the Incat Tasmania shipyard in 1999 and was operated by the Australian Navy as a troop carrier. She was in military service until 2001.

HMAS Jervis Bay was extraordinarily successful. She transported troops to trouble spots in New Guinea and Indonesia. As an auxiliary ship, her high speed of 38 knots enabled her to prove her worth in numerous maneuvers, even before American eyes. The latter were very impressed by the large transport volume, seaworthiness and long range of this ship. Although the fast military ships such as the speedboats, light destroyers and others had so far tended to be characterized by a short cruising range, this rule no longer applied. Why?

The progress that has taken place in aluminum shipbuilding now makes it possible to build hull structures that were unimaginable some 30 years ago. Increased material qualities, much better welding techniques and computer-aided calculation of ship structures led to the construction of the first large fast ferries for the canal service, Incat Tasmania's 74-meter catamarans, in the mid-1980s. The small shipyard on the Australian island of Tasmania had hit the big time in the market with an order for two of these high-speed vessels. The „Hoverspeed Great Brittain" proved its efficiency by winning the „Blue Ribbon" in 1989 for the fastest Atlantic crossing at that time from New York to Southampton. Although purists to this day are of the opinion that she is not a true passenger ship and therefore does not deserve the award, 36.97 knots as the average speed for this distance speaks for itself. This speed was achieved without any modifications to the ship during a simple transfer voyage. The introduction of powerful and light diesel engines began in the German S-boats during the world war and continued globally in almost all new fast boats built after the war. The advances made in this process paid off with high interest rates for the fast ferry industry that emerged in the 1980s.

Since civil operators had considerable reservations about the gas turbine as a means of propulsion, shipbuilders had to switch to diesel drives. These were still repairable for operators accustomed to conventional technology and were also more fuel-efficient than turbines at the time. Outputs increased rapidly from 1,000 to as much as 9,900 hp. On the propulsion side, the new waterjets very quickly gained the upper hand, These propulsion systems, based on the recoil principle, can be easily installed in slender hulls and are increasingly more efficient than propeller drives at speeds higher than 30 knots. They can also be used for steering.

„Jets" are now built in a wide range of sizes. They are mechanically simpler in design than propeller units with long and rapidly rotating shafts. From small units only a few inches in diameter to jet openings through which a truck could pass without hitting it, all sizes are represented. The largest water jet built to date is such a monstrous unit, designed to produce about 40,000 horsepower. The ship to match it has yet to be built. The high control capability of the jet propulsion system is based on swivel nozzles and reverse thrust flaps that deflect or redirect the ejected water jet forward. This does not require the machine direction to be reversed. The high responsiveness makes jets an indispensable propulsion system for high-speed ships. The final and perhaps decisive factor in the superior performance of modern high-speed ferries has been the advances in hydrodynamics made primarily in the last two decades. Today's catamarans, as well as trimarans, have long left behind the level of the old planning and semi-planning hulls.

Essentially, two concepts have prevailed in the world of large fast ships that primarily define how a fast ship copes with the sea state it encounters. Conventional seagoing vessels pitch, roll and yaw. This lists all the effects that make it so difficult for seagoing vessels to move.

The individual movements are determined by several force components, such as gravity, inertia and the ship's buoyancy. Gravity tries to pull a ship as far as possible into a wave trough. It is directly counteracted by buoyancy, which forces the hull back up - most of the time, anyway. Inertia is opportunistic, acting in concert with both gravity and buoyancy. Once the heavy hull gets going, it does not simply stop its motion at the onset of the counter component - that is, lift.

*The USS „Spearhead" was part of the US Army fleet, not of the US-Navy! She was equipped with a swiveling universal ramp at the stern to be able to put down vehicles up to the size of a heavy tank on any kind of wharf.*

Inertia causes the hull to sink a little deeper into a wave trough than it should or to rise a little higher than just the buoyancy should do. The designer can get this mesh of different forces under control in slow craft simply by shaping a suitable hull, especially at the forecastle. But this no longer works for fast ships: The short reaction time required to regulate the hull behavior through buoyancy is no longer achievable to ensure smooth sailing. Therefore, driving a planing boat through a rough sea is like racing on very rough cobblestones. The shocks given to the hull by the waves can knock standing passengers off their feet.

Even the first fast catamarans were not spared this phenomenon. Although they no longer rolled as badly as slim monohulls, much to the chagrin of those on board, they still jumped through the waves quite lively.

The Australian designer Phil Hercus came up with a solution in the mid-1980s: He drew a catamaran that would pierce the waves with its bow tips rather than cross them - the „wave piercer" was born. The first two examples of this type did not look very convincing to the observer, but they already worked quite successfully in rough seas. Phil Hercus had transformed the catamaran hulls into flat pontoons connected by slender sup-

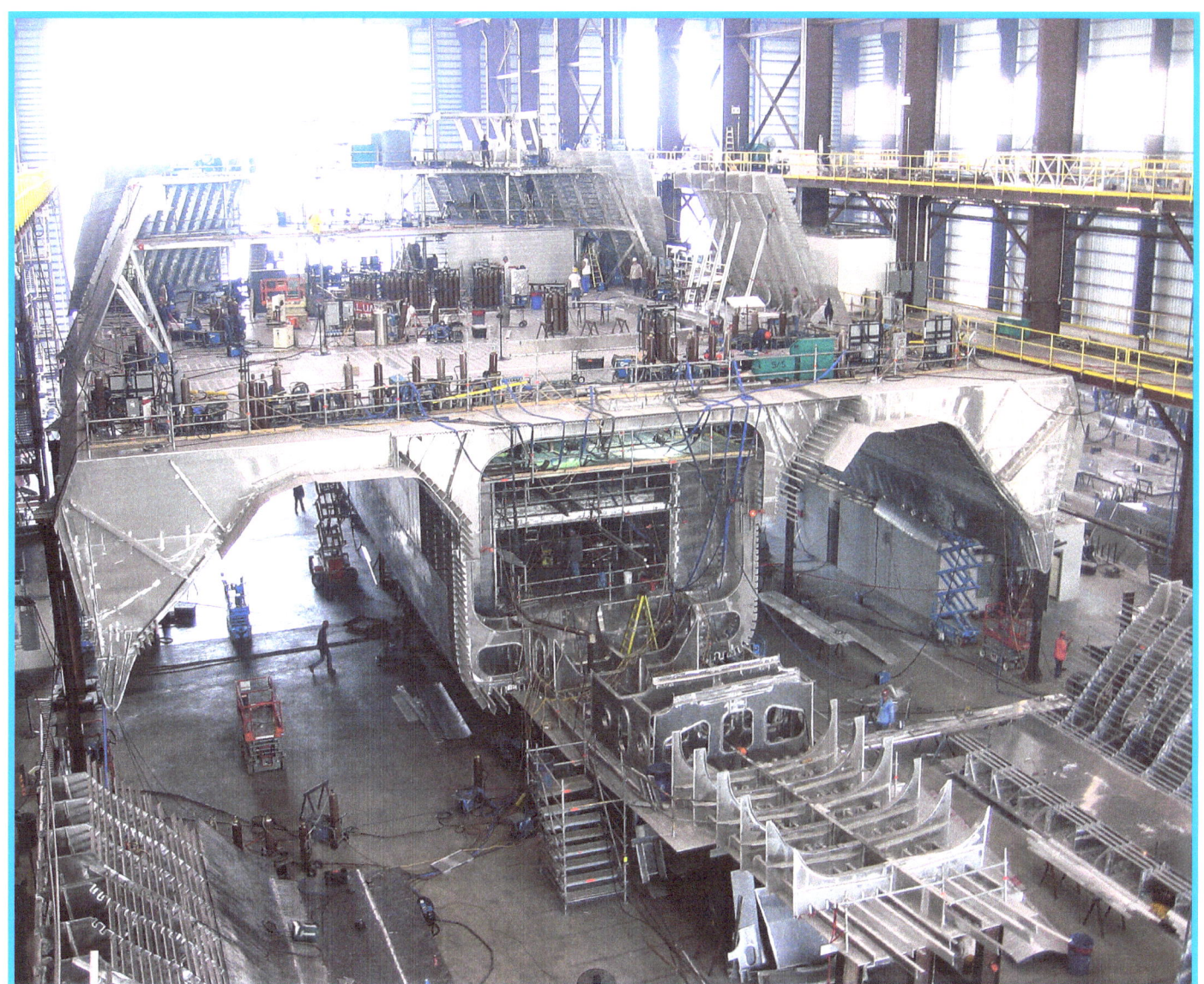

*(Above) Here you can see the hull of a large trimaran in the shell. Aluminium is a light but less strong material, which means that the spacing between frames is smaller than in steel shipbuilding. The outer skin of this ship is welded onto a network of frames and longitudinal struts. Together they form a strong shell structure that can withstand all kinds of loads. The trimaran under construction here is the Litorial Combat Ship USS „Independence" (LCS).*

ports to the central hull, which was some height above.

This allowed higher waves to rush over the catamaran hulls without encountering resistance and did not bother the rest of the ship. But access to the engines in the two hulls was impossible during the voyage. Likewise, the props were too weak to support heavier loads on the even larger catamarans planned.

Hercus tirelessly tested new concepts until the current shape was established by merging the props into a slightly angled sidewall and shaping the bow tips into a sleek „mouth."

A large civil order for the fledgling Incat Tasmania Ltd. shipyard from the English shipping company „Sea Speed" for two 74-meter catamarans was the start of a major Australian success story. Incat Tasmania Ltd. is today one of the two largest aluminum shipyards in Australia. The last examples at the time of writing, launched at Hobart on the island of Tasmania, were two 112-meter vessels for a Japanese owner. A third vessel of this length was planned for the US Navy, the so-called „HSV" (somewhat unimaginatively meaning „High Speed Vessel"). Catamarans of 150 meters in length are already being planned. They will be able to carry over 2,000 passengers and many hundreds of cars.

The aforementioned „HMAS Jervis Bay" was leased from the Australian Navy on the initiative of forward-looking officers. On several occasions, HMAS Jervis Bay was able to demonstrate its superiority as a transporter for vehicles and troops resulting from its speed and capacity.

This attracted the attention of the US Navy, which had been looking for solutions to the „slowness" of conventional transport ships since the 1970s. Large-scale studies of large hovercraft and hydrofoil frigates had been undertaken, but the extreme cost of these technologies had made them unsuccessful. But now the Navy was presented with a surprisingly simple and affordable solution on a silver platter, and all it had to do was grab it. In contrast to the Cold War, the focus was no longer on submarine hunters or missile destroyers. transport ships were planned that could cost-effectively move troops and their equipment quickly and globally. True to the statement of a general of II. World War II that the real weapons are not the guns, tanks and cannons, but the soldiers, things have changed a lot during the war on terror.

For some time, it was believed that the time of the great field battles, of the loss-laden wars of position and of the great naval battles was finally over. However, the war that broke out in Ukraine in 2022 has taught the world better. For this very reason, it is very important in the Western community to be able to provide an effective deterrent not only on the side of weapons. The ability to move troops around the world extremely quickly is also an equally effective „weapon." Since the time of World War II, the U.S. armed forces have been paying as much attention to the maintenance of their transportation capabilities as they have to the condition of their armed units.

The transport aircraft is the most important factor in this response capability in just a few days, but the U.S. Navy and U.S. Army primarily maintain very large fleets of mostly second-hand RoRo and container transport ships. They are the ones that, when in doubt, transport the large numbers of vehicles and equipment by sea to the Asian or European theater that would be needed by the U.S. Army in a theater of war. Even mobile port facilities are part of the planned equipment. In most cases, however, today's common warfare is what is commonly called limited warfare. Actions against pirates, guerrillas or drug traffickers are commonplace today. Much of today's warfare takes place undetected by the public. On land, this type of warfare is carried out by specially tailored units that, equipped with relatively light equipment, can be deployed quickly over global distances.

Meanwhile, limited conflicts are even being „outsourced," with so-called „security companies" doing the dirty work on the ground. High technical capability of weapons and vehicles, surprise appearance on the scene, excellent training and great professionalism characterize the soldiers of the Rangers, Special Forces, Marines, Royal Marine Corps and Légion Étrangère (Foreign Legion) of various Western nations. Most major nation-states have such units, whose composition and armament are often classified. In Germany, the "*Kommando Spezialkräfte*" (German for "Special Forces") combines the capabilities of combat swimmers and the corresponding specialists of the army to form an internationally recognized special unit. In the meantime, the German Air Force has the A-400 M large transport aircraft, a transport vehicle that meets their requirements. The Navy, on the other hand, is currently hardly in a position to offer floating bases of operations. In the German Navy the Special Forces utilizes the very silent modern German Submarines as a basis for clandestine operations.

But it is now planning to build large multipurpose combat ships to fill this gap. The United States now has the most capable and largest transport air fleet ever. The balanced mix of C-130 Hercules, C-17 Globemaster II and C-5 Galaxy tactical transports enables the U.S. Air Force to deploy heavy equipment as well as soldiers operating lightly around the world

*(Above) This large jet actuator shows the core elements of its type: The nozzle that can be swiveled to the side, a deflector flap that can be swiveled into the water jet for the reverse thrust, and the attachment to the water inlet on the right side.*

within about 12 hours. In addition, there are chartered cargo and passenger aircraft.

However, even this large air fleet has its limits, especially when it comes to transporting tanks, heavy guns or air defense systems. It is very costly to have each heavy transport aircraft carry a tank or a self-propelled gun. However, light weapons are not usable on all occasions. Similarly, the value of light armored vehicles in a particularly hostile environment is questionable.

Often it is not possible to survive without heavy equipment if the survival of one's own troops is to be ensured. This is the moment when the transport ship is still superior to the fast aircraft. A prime example are the seven fastest container ships ever built of the type „SL-7", which was also built in Europe for the shipping company Sealand in the 1970s. The original full container ships have been converted into RoRo multi-purpose vessels and are at rest in various US ports. They can be made ready for departure within 96 hours and then race to the world's trouble spots at speeds of almost 32 knots.

But these ships are extremely expensive to operate, burning about 340 tons of fuel oil per day. In addition, they are massive ships that are too valuable and too large for use in „hot" combat zones. They are also not flexible enough to perform tasks other than transporting heavy equipment. As a solution for limited warfare, where, for example, commandos need to be carried in company strength, supplies need to be transported quickly to focal points, or relief supplies need to be carried to the scene of natural disasters, the new catamarans appeared to be much better suited.

Lightly built, fairly seaworthy, very fast, and with a long range, they can carry many times the cargo of a C-5 Galaxy across the Atlantic or another ocean in a very short time. At the same time, operating costs are much lower. And you can buy the equivalent of a single one of the heavier airlifters like the C-17 or the C-5, you can buy several of these ships.

So it was decided to lease and test three such ships for a few years. Two of these catamarans were from Incat Tasmania

("Joint Venture" and "Spearhead") and a third from Austal Ships ("WestPac Express"). But after the return of the three test vessels, things got serious: a design for a high-speed transport vessel over one hundred meters long for all branches of the armed forces was put out to tender.

The aforementioned companies Incat Tasmania Ltd. and Austal Ships PTY participated in the competition via partner companies in the USA. This is because an old law known as the Jones Act prohibits the import of new ships into the USA. If a foreign company wants to gain a foothold there, it is forced to set up a subsidiary there or join forces with sales partners. Some American politicians would prefer to extend this protectionism to all industrial goods, but fortunately the American upper class does not want to do without luxury automobiles imported from Europe.

Austal Ship won the tender at the end of 2008. Austal's ships follow a fundamentally different design principle than those of Incat. They prefer the so-called "semi-SWATH," a catamaran hull form based on a slender, quite deeply dipping underwater hull with rounded curved lines that gently merge into one another. The bow tips are given a kind of bulbous bow. Due to the larger draft, buoyancy is gained from the so-called molded buoyancy. Because of their lower drag, these catamarans, known as "semi-SWATHs," can reach high speeds with less energy input. Precisely planned shaping of the transition areas between the underwater hull and the upper hull, together with the use of T-foil systems, can reduce the effect of the swell in the same way as wave piercers. Austal and Incat are in fierce competition with each other and the loss of the JHSV contract was certainly a major blow to Tasmanians.

*(following page) A deck plan of the "TSV-1X Spearhead". The ship is one of the largest catamarans built, measuring almost 98 meters in length. It is made entirely of aluminum and was fully welded together. Four robust Ruston "high-speed" engines propel it to 38 knots fully loaded. Its sister ships were allowed to operate as civilian ferries while the Spearhead was required to serve her military duty. The U.S. Army was very impressed with the catamaran's versatility and flexibility. It was tested as a minelayer, transport ship and command carrier. The "X" in its name indicates the traditional designation in the U.S. Armed Forces for test patterns. TSV-1X had been leased between 2002 and 2006. After (honorable) discharge from the Army, the ship was converted to a fast ferry and sold.*

*The Spearhead is a typical wave piercer, which is evident from the forward bow tips. The central bow carries the ship over too large waves. At the same time it prevents a hard impact in such high seas. The colorings of the decks indicate their use: The yellow deck is an area for additional troops, who are passengers on board. The greenish area is the crew deck with small cabins. White or light blue are messes. The turquoise area is the "Command Section" with the bridge at the top and some additional rooms, such as an office for commanding troops on board.*

*Aft is a movable vehicle ramp that can be moved to the side or aft. In front of it, the vehicle deck can be seen. A landing platform for helicopters up to the size of the Sikorsky UH-60 "Blackhawk," the standard helicopter of the armed forces, has been installed on the aft upper deck. There is no hangar.*

*(Below) On 1st October 2016 Houthi rebells attacked the HSV-2 "Swift" with RPG rockets or an anti-ship missely and damaged the "Swift" very heavily at the foreship. No casualities have been reported.*

*The terrorists claimed to have sunk her, but the vessel was resqued by the US Navy and towed to Eritrea. Later the vessel was transported to shipyard and have been repared. The ship later was sold to a Greek ferry company.*

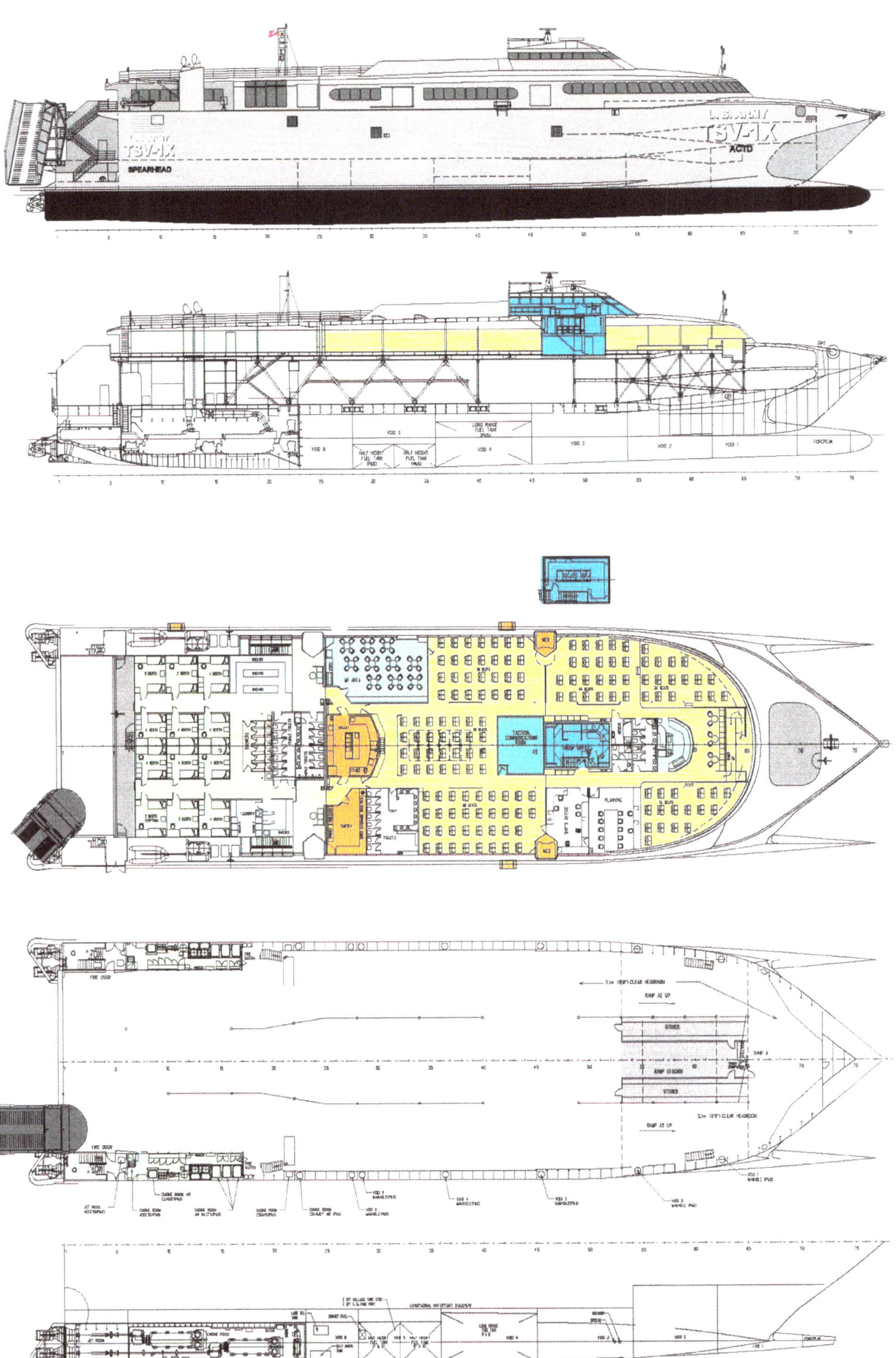

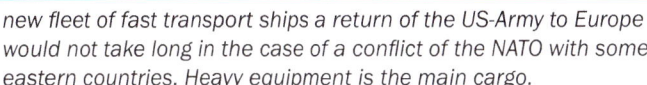

(Above) A US Army 'Striker' APC rolls up the ramp of a high-speed catamaran. Loading such a vessel only takes up to 3 hours. This gives the Army a new kind of quick reaction capability. The atlantic crossing of such a fast catamaran takes just three days. With the new fleet of fast transport ships a return of the US-Army to Europe would not take long in the case of a conflict of the NATO with some eastern countries. Heavy equipment is the main cargo.

(Left) A unit of the new JSHV class under sonstruction in an US shipyard. The so called "Semi-SWATH" form could be seen the best at the crosssection of the bow on the right side of the picture. This specific hullform ensures a high seaworhiness for the vessel.

*HSV-X1 - The USS 'Joint Venture' an Incat 96m-catamaran, which has been built originally for truck transportation on the Bass Strait between Australia and Tasmania. It was leased by the US-forces from 2001 to 2006. After the military service she operated under civil flag in Europe.*

# The winner in the end

*(Above) The winner of the JHSV tender at the end of 2008 was the Australian company Austal Ships from Henderson on the West Australian coast. The „Joint High Speed Vessel" will be a multi-functional high-speed transporter carrying troops and equipment for the U.S. Marines, Army and Navy.*

The goal is to be able to transport rapidly deployable reaction forces to any coast in the world within a few days. To do this, the U.S. will station up to nine of these vehicles at strategic points that allow short access times, such as Diego Garcia in the Indian Ocean. From there, any point on Asian or Arab coasts can be reached within 2 to 3 days at an average speed of about 35 knots.

Although the U.S. operates the largest military transport air fleet in the world, it has been found that it quickly reaches its performance limits when deploying ground troops. In fact, a C-17 can only carry a single M1 main battle tank. In contrast, a JHSV can carry 635 tons of cargo, including 10 to 15 tanks and up to 364 soldiers.

Since even the U.S. can't simply deploy over a hundred large aircraft quickly to an airfield, the JHSV's ease of loading makes it the superior mode of transportation. It can be fully manned and loaded in just a few hours, as civilian shipping exemplifies with car ferries. It consumes less fuel and is much cheaper to procure as a single ship unit than a C-17 costing over $250 million. The JHSV is one of the largest aluminum-built ships in the world:

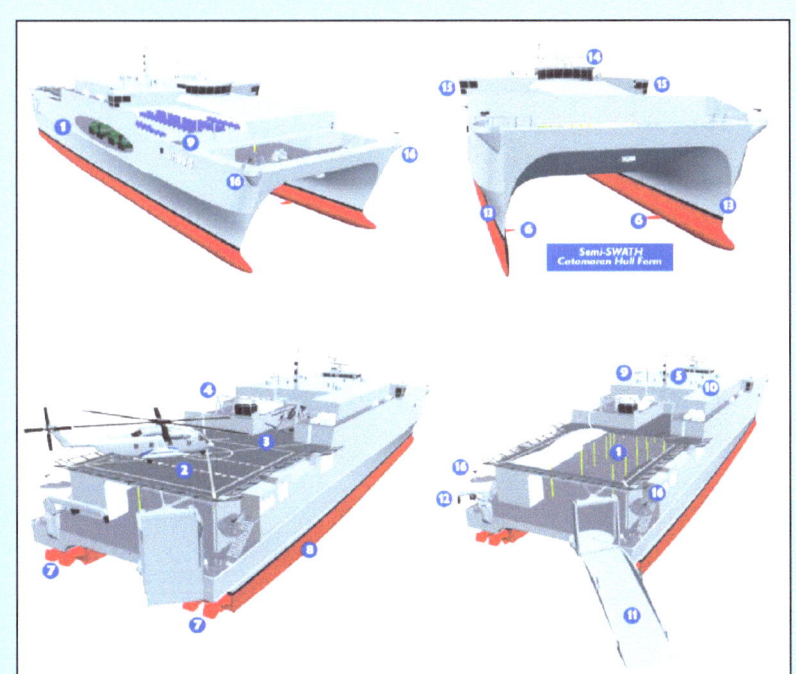

**Key data:**

| | |
|---|---|
| Length: | 103.0 m |
| Width: | 28.5 m |
| Draft: | 3.83 m |
| Propulsion: | 4 x MTU Diesel 20V8000 (9,100 kW / 12,200 hp each) |
| Speed: | 35 kn / max. 43 kn |
| Range: Loaded: | 1,200 SM |
| Empty: | 5,600 SM |
| Cargo volume: | 1,863 cbm |
| Troops: | 341 |
| Crew: | 41+48 |

(1) Vehicle deck; (2) Helicopter landing deck; (3) Stowed helicopter; (4) Flight Control; (5) Bridge; (6) T-fin; (7) Jet drives; (8) Hull; (9) Crew living deck; (10) Side gangway; (11) Vehicle ramp; (12) Auxiliary crane; (13) Semi-SWATH hull; (14) Bridge; (15) Side operator stands; (16) Side troop quarters

*(Above) T-AKR 292 Regelus. The large ship is packed with tanks, guns and other various vehicles needed by an U.S. Army combat regiment in action. The associated soldiers arrive for deployment by air on chartered airline aircraft.*

*(Above) The JSHV's „main weapon" unfolds - the vehicle ramp.*

*(Above) A JSHV from the front. The big catamaran tunnel makes the ship resistant to most common waves*

*(Above) A new U.S. Army logistics catamaran under construction at Austal USA in Alabama. .Although it is unarmed, its cargo is also deadly: soldiers, tanks and guns - along with tons of donuts, pizzas and hamburgers.*

*(Above) The USS "Trenton" (JSHV-5) beeing near to its launching in March 2015.*

*(Below) The Tasmanian contribution to the JHSV program, a 112-meter 'wave piercer'. The modular hull has already been built twice in civilian version for Japanese ferry operators.*

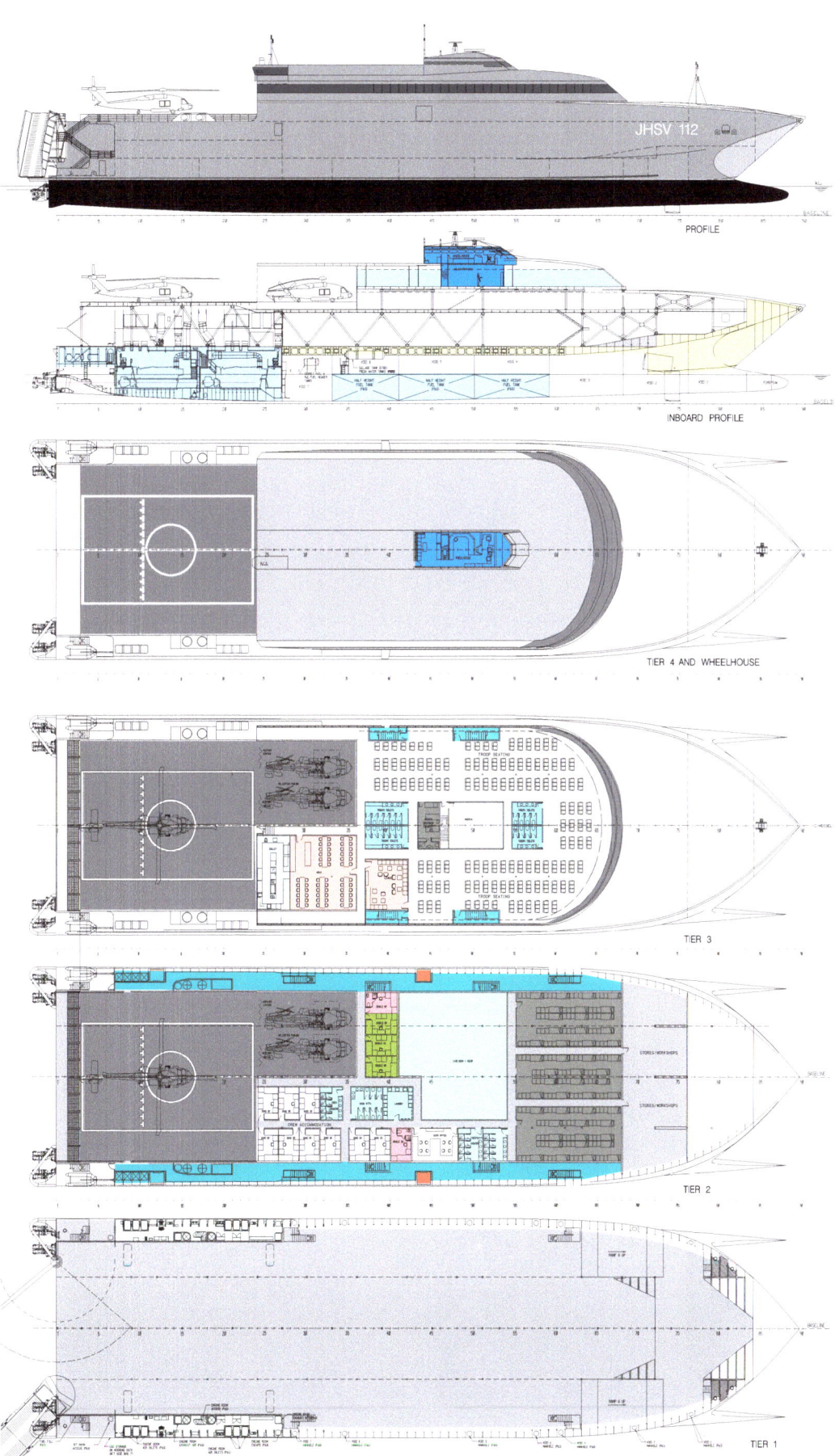

# Ships as from another star

The desire to be able to hide from radar is almost as old as this tracking technology itself. As early as World War II, the Germans wrapped the periscopes and snorkels of their submarines with so-called „Wesch mats". These materials contained iron filings and were designed to absorb the incident radar beams instead of reflecting them back. The absorbed energy was converted into heat. But the shape of a body also came to the attention of experts when the first flying-wing aircraft of the early 1950s were found to be difficult to detect by radar. „Stealth" research by the U.S. and indeed all other nations was quickly covered with the cloak of secrecy.

The result was secret test flights of strange-looking flying objects beginning in the early 1980s. Curiously, these aircraft were mistaken by many uninitiated citizens for UFO sightings when they once howled over their heads at night. But no matter how otherworldly these aircraft looked, the F-117 was no longer considered particularly fun by the Iraqi military by 1992 at the latest, when it began taking out heavily defended targets in the core of Baghdad. What works in the air also works on the water. At Lookheed-Martin, they dreamed up the „Sea Shadow," which looks as little like a ship as you can imagine. But this strange structure, composed of triangular shapes, sailed around at night off the Pacific coast of the USA, not visible by RADAR. Meanwhile, it has become a trend to give the new warships a certain anti-radar shape on the way. It is characterized mainly by the fact that no outer surface is constructed at right angles to the water surface anymore. Instead, they are angled either slightly toward the water surface or upward.

Likewise, the trend is to no longer use curved surfaces, especially on superstructures. Lattice masts and round components are also out of fashion. Today's warships are extremely angular and smoothly shaped on the outside, and no longer stare from the most diverse antennas, as they did in the last century. By angling flat surfaces toward the water surface or sky, an incident radar beam is reflected back to one side only, but not to the enemy.

On the other hand, round bodies reflect the radiation in every direction. On the water it is actually very easy to predict from which directions enemy radar beams are to be expected, namely

*(Below) The weared looking STEALTH experimental ship „Sea Shadow" from Lockheed.*

always more or less parallel to the water surface. Therefore, it was not very complicated to design appropriate ship shapes. It was certainly more difficult to convince the clients of the new non-traditional shapes. The stealth shape was supported by radar-absorbing paint coatings or coatings containing iron powder and chips like the old „Wesch mats". Their composition today is very refined and top secret.

On many battleships, they have even gone so far as to make fittings and deck gear either anti-radar in shape, or retractable. Dinghies are hidden in garages and even ladder steps are given a prism-like cross-section. In some cases, these measures reduce radar backscatter to about 25 percent of that of older shipboard units. For the computer of a guided missile, this can pose a major problem when it has to distinguish between several targets radiating back unequally. For example, it is likely to mistake a decoy that has particularly strong backscatter for the actual target. Today's designers are consistent enough to attenuate infrared back radiation as well. Smoking and spark-spewing vents are finally a thing of the past on modern combat ships. But one thing bothered Swedish planners about the new designs when they were planning a new corvette for the Swedish Navy: The construction material of conventional combat ships is still metallic! Whether steel or aluminum - it still radiates back better than other materials.

Therefore, the idea was to construct a plastic hull for the new Visby class. So the ship was glued together from multilayer plastic plates. These panels consist of glass-fiber-reinforced synthetic resin outer skins encasing a layer of ‚rigid foam or of PVC honeycomb cores. These „sandwich panels" are incredibly strong and lightweight, However, they can only be glued together. Welding and other joining techniques are not applicable. Once the Navy approved this unusual concept, they went to work building the type ship, which set new standards when it was completed. It was built to be as consistently radar-repellent as could be justified. In many places, even the railings were omitted, because they could increase the reflectivity. All weapons have been hidden below deck with the exception of the gun turret. Even the gun barrel is retracted when not firing.

The Visby does not know propellers. She is accelerated to the minimum 35 knots she reaches by gas turbines acting on jet engines. Long distances are covered at about 15 knots by additional diesel engines. She can strike hard because she is equipped with eight sea target missiles with a range of 200 km. Other weapons include anti-submarine torpedoes and small underwater robots that can be used for mine countermeasures. Complementing this armament is a fully automatic 57mm gun turret from Bofors of Sweden, which like everything else on the Visby has a radar-deflecting design.

The rotational speed of this turret and the rate at which it moves the gun barrel is so fast that a human could hardly avoid it - the gun barrel would strike him dead. The weapon system can take out all flying objects, including approaching missiles, within a few kilometers distance without human intervention. Nevertheless, it can also attack ground targets at a distance of 17 km. An automatic ammunition change system allows it to react extremely quickly to different targets. Since the gun is lighter

**UFO alert?** *These two corvettes of the Swedish Navy look as if they are of extraterrestrial origin. They are the first two units of the new Visby class, which have caused quite a stir internationally.*

*(Above) Visby corvettes, more than any other ship before, were consistently designed for protection against detection. Wherever possible, equipment and appliances have been relegated under the hull. Helicopters can land and take off aboard, but the hangar is still missing. The only weakness compared to other corvettes is the lack of a more effective anti-aircraft system. Nevertheless, this rather small ship is versatile and can do a lot of very different tasks together with sea surveillance and submarine hunting to mine sweeping.*

and more modern, it is now often preferred to the already familiar OTO-Melara 76 mm gun. It is even considered as armament for frigates. Compared to the firepower of a World War II destroyer, such a new 57mm turret replaces the entire anti-aircraft armament of an old ship. The modern gun can fire shells with an impact fuse, an armor-piercing effect, or a radar standoff fuse.

This is something that laymen often fail to realize: Today, the intelligence of weapons systems determines success. Computers are replacing human labor, and are many times more efficient. It's no wonder that a modern missile fast boat in a direct duel could put one of the old heavy 20.3 cm cruisers out of action before it would get the boat in sight. Today, the main objective is no longer to destroy an opponent, but to disrupt his weapon systems and drives in such a way that he fails as an opponent. Reaction times are extremely short and the firepower of today's small calibers is driven to better levels than was even conceivable 60 years ago by highly effective propellants and explosives. Stealth technology, balanced armament and equipment, and powerful and quiet propulsion make the ship an adversary on a par with far larger and more expensive destroyers and frigates of other navies. Onboard helicopters and anti-aircraft missiles are not needed in Sweden, as the ships always move within range of the Swedish Air Force.

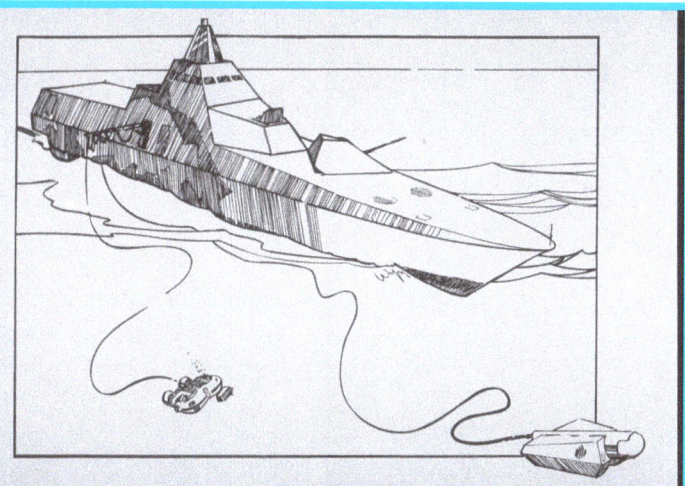

*(Above) Hand sketch of the method of operation for mine detection. The underwater robot (ROV) on the right searches for mines, while the one on the left places explosive charges at the mines it finds. Divers are no longer needed with this technology.*

(Below) In modern navies, saving personnel through automation is increasingly important. Three or four highly qualified officers and NCO make up the corvette's bridge crew. There is no longer a telephone operator, radio operator or machine telegrapher. Even the coffee is now fetched by everyone themselves.

The Visby has windows only on the bridge. Therefore, the crew has to get used to seeing the sun only during off-duty hours - if they are allowed on deck. But the quarters are comfortable enough to avoid the atmosphere of an old submarine.

# The anatomy of the "Visby"

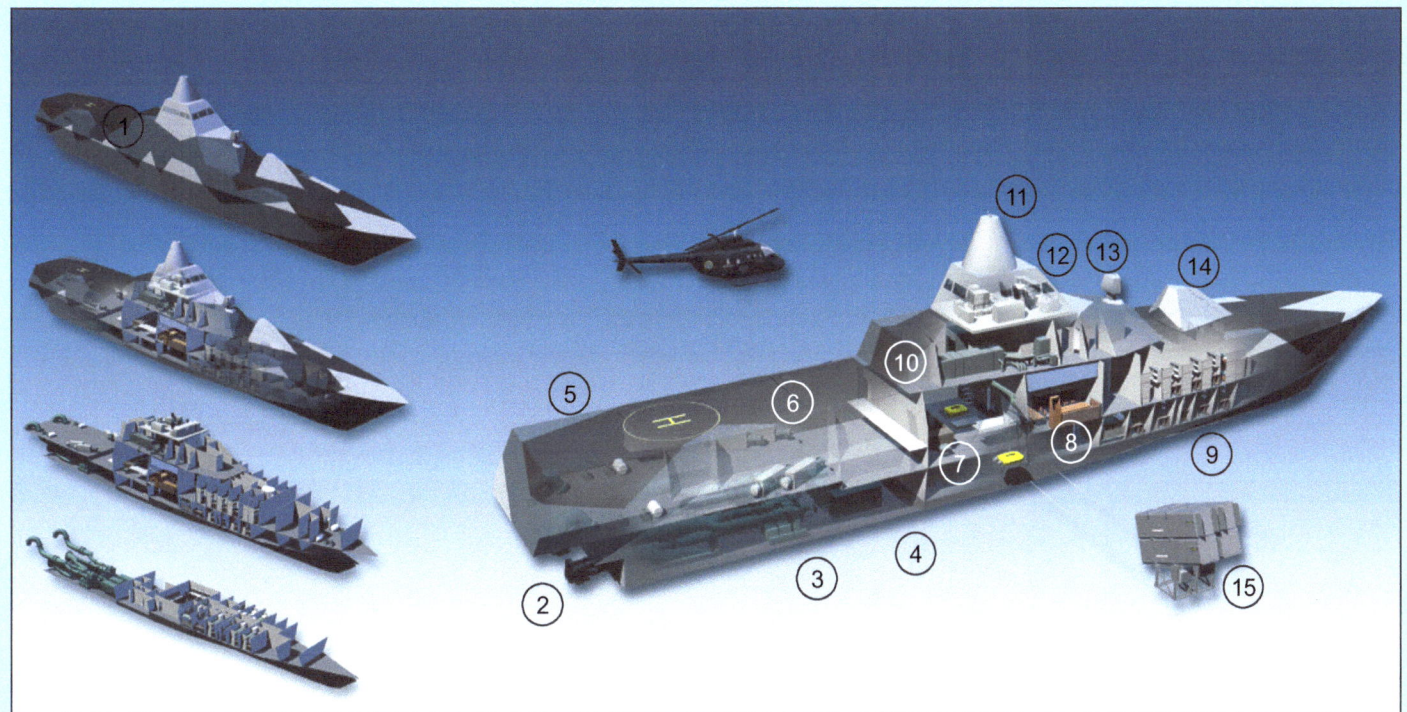

1. The ‚Visby' in the various stages of „undressing". The rather small ship nevertheless has a large interior and can accommodate a large amount of equipment and weapons.
2. The water jet drive, which provides propulsion and control. Above it are the exhaust pipes of the gas turbines, which were directed at the water jet. Thus, the exhaust gases are cooled to suppress the infrared locatability.
3. The four combined 16,000 kW TF-50 gas turbines from Honeywell. As compact, lightweight and powerful units, gas turbines are superior to diesel engines.
4. Two MTU 2000 diesel engines with 16 cylinders of 2,600 kW each serve as drive for slow speed stages.
5. The helicopter landing deck. A hangar can be retrofitted. A towable sonar cable is also housed here.
6. Optional installation space for the helicopter hangar. There are also two anti-submarine torpedo tubes on each side. Eight RBS-15 anti-submarine warfare launchers were also installed in this area.
7. Here, two underwater robots (ROV) for mine countermeasures and a motorized inflatable boat can be deployed by cranes extending sideways.
8. The Combat Information Center' (CIC): This is where all information from the sensors and data links to other units converge
9. Crew accommodation area.
10. Space for auxiliary equipment and the air conditioner. The interior can be fully protected against toxins and nuclear contamination
11. Under the angular „sugar loaf" are the radar antennas and the receiving antennas for radio and electronic radiation from other units.
12. The navigation bridge from which the ship is driven. In case of battle, however, it is commanded from the CIC (See there).
13. Optical sighting system for the 57mm gun. The system can also see at night via residual light amplification.
14. The fast firing 57mm gun. Due to its precision and firing rate of 220 rounds per minute, it achieves several times the firepower of a large World War II destroyer. It can also repel incoming sea target missiles. Its firing range is up to 17 km.
15. The RBS-15 missile system fires a missile capable of driving a 200 kg armor-piercing warhead into a ship's hull while flying low at about 1,070 km/h with a range of 200 km. It is currently considered the most powerful conventional anti-ship missile weapon in the West.

Overall, there seems to be nothing that the Visby cannot do. Her detectability is very low. In rough seas, she can only be detected by radar at a distance of 13 km, and using electronic own combat control from 11 km. This is then much too late for any opponent.

The internal design of a STEALTH combat ship is dictated by the need to hide all essential systems and equipment inside. The smooth outer hull must not be interrupted by anything that would reflect radar beams. In addition, in today's world, systems must be easily replaceable with new ones. It is therefore best if they are pre-assembled in handy boxes or containers. The ship then spends only as much time in a shipyard as is needed to replace the containers. Today's combat ships are far too few in number and too expensive to afford a long shipyard stay, as was necessary not too long ago.

The Visby is one of the largest ships ever built entirely of plastics. Fiberglass scrims joined by synthetic resins were used. Instead of being built in a prefabricated mold, however, the hull of the Visby was glued together from cut-to-size sheets of this material were glued together. Despite the prophecies of doom of all skeptics, this resulted in a fireproof, very solidly built and seaworthy ship. The plastic is lighter than steel or aluminum, allowing for better sailing performance. The hull will never rust, saving Swedish taxpayers a lot of money on ship maintenance. The Visby contains far fewer metal components than conventional ships, which greatly reduces detectability by radar. The magnetic field changes caused by the ship are also very small.

The chance that the magnetic fuses of mines or torpedoes will respond to the Visby is much lower than with steel hulls. These corvettes have brought a very large edge to the world of combat ships, the impact of which has not yet been fully appreciated. „Invisible" ships will dominate the sea in the future, surprisingly appearing off the coasts of their opponents.

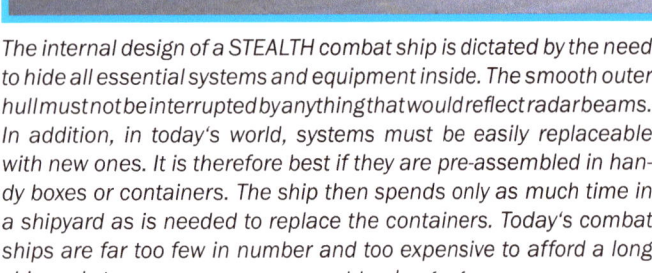

*(Above) Norway's counterpart to the Visby class is the ‚Skjold'. Also constructed of plastics as a stealth ship, it nevertheless represents a different technical concept. Her hull is not a semi-planning one, but more close to that of hovercrafts. Air is blown into the cave between two catamaran hulls, held together by skirts at the bow and stern. The air cushion reduces the draft to such an extent that the Skjold can sail faster than other vessels with less effort. The Skjold is smaller than the Visby and more in the class of fast combat boats.*

In Norway, too, defending its own coasts is a key task for the navy. The special shape of Norway's coastline, with its winding fjords and bays, makes it difficult to deploy destroyers. The German navy had to learn this lesson as early as 1941 when it conquered the northern country.

In many waters, there is simply not enough room to maneuver larger ships at high speed. The Norwegian Navy therefore developed an impressive number of speedboat classes and used them as combat assets in coastal areas. Defense of the maritime areas off the coast was left to NATO, which, fittingly, had placed one of its main focal points nearby in combating the northern fleet of the former Soviet Union. In addition, Norway's air forces can be much more effective against formations approaching the coast because it is exceptionally long. It would take too much time for a battleship force to approach the scene. In contrast, fast-attack flotillas can be cost-effectively dispersed along the coast in camouflaged positions and engage the enemy in raid-like „raids." Norway even developed its own compact sea target missile system that can be fired from the fast boats and from aircraft, the „Penguin."

It has a shorter range than the American „Harpoon" or the French „Exocet," but this is not so essential in the fjords. It is fired reactively, like an aerial torpedo, in some cases directly upon enemy sighting. An older „Storm-class" fast craft can carry six of these missiles, giving it considerable combat power. A 76 mm turret sourced from Italy by OTO adds to the mix of weapons.

The new Skjold boats use radar beam deflecting and absorbing materials. An extremely high speed of 60 knots (111 km/h) can be achieved in calm seas. In bad weather, the speed is reduced to 45 knots. Two gas turbines, each with 8,160 hp, provide fast cruising, while two 1,000-hp diesels act in slower passages. The design as a hovercraft with fixed sidewalls is called an ‚SES' or ‚Surface Effect Ship'. SES are highly resistant to detection by submarines and acoustic mines. The layer of air between the hull and the water surface prevents transmission of engine noise into the water. The frictional resistance of air on water is much lower than that of water on the hull, which explains the high speed despite the rather low propulsive power.

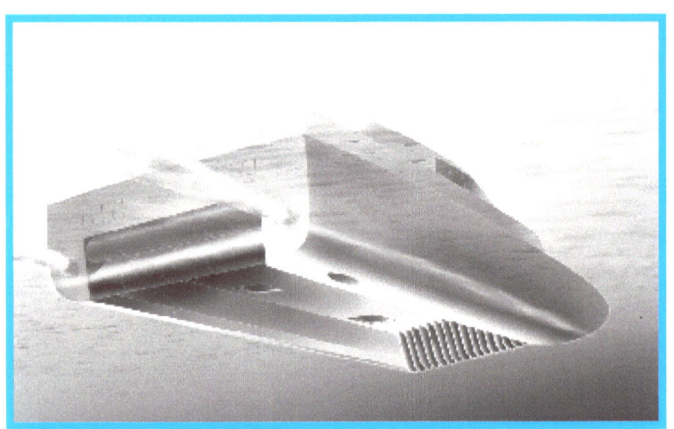

*(Left) The fuselage of a sidewall air cushion (SES) vehicle from below. Visible is the cavity containing the air bubble on which it moves. The flexible skirts at the bow and stern hold the air bladder together even when passing through waves. The sidewalls dip a little into the water, but not to the point where the drag would become as strong as that of another type of boat. This allows an SES to reach high speeds - sometimes up to 50 knots.*

*(Below) The forerunners of the Skjold class were the 'Alta' class minesweepers from Norway. These GRP-built ships were the first mass-produced SES ships in a fleet. They were deployed with great success beeing less vulnerable to the most mines.*

The Skjold, unlike earlier SES, is designed to operate effectivly also at slow speeds. While conventional SES began to become quite uncomfortable and uneconomical in this mode of operation, the Skjold is a cross between a water displacing catamaran and a fixed sidewall air cushion craft (SES). The air cushion only reduces the draught to such an extent that the water resistance is considerably reduced, without losing the characteristics of a real ship. Thus, she is still very seaworthy even in high seas. The US Navy was allowed to test the Skjold for a few weeks and was very impressed.

| | |
|---|---|
| Length. | 47,5 m |
| Width: | 13,5 m |
| Draft: | 2.3 m |
| Propulsion: | 12,000 kW / 16,092 hp |
| Speed: | up to 60 kts |

# America's invisible warriors

*(Above and overleaf) The full extent of the innovations on the LCS is revealed by a look inside the building dock. The trimaran hull form, never before used on a warship, is said to offer significantly reduced water resistance and a better ability to sail very quickly through high seas than conventional designs. The LCS, along with a civilian ferry from the same manufacturer, is the largest trimaran built to date.*

As many critics claimed in the 1990s, the U.S. only started planning new types of combat ships at a rather late stage. But it makes sense, given the scale of U.S. defense planning, not to take up the first technical innovation that comes along.

In fact, intensive work on projects for alternative combat ships has been going on since the seventies. The weaknesses of conventional missile destroyers and frigates were already well known at that time. Insufficient speed, ineffective weapons, visibility on radar, and exposure to submarines are some of these factors that make conventional combat ships good targets on the surface today. A Lookheed P-3C maritime patrol aircraft can carry up to four Harpoon missiles on its wings, and a long-range B-52H bomber can carry more than 12 in its belly. So a small number of those bombers could wipe away a complete fleet of destroyers and frigates from the surface. without being threatened by those warships.

In fact, most fighter aircraft in service today can be equipped with sea target weapons and hunt over water. Satellites and radar networks help track the targets. Today, the sea is dominated from the air. Aircraft can be deployed anywhere, are available in large numbers, and require little travel time even to remote areas of operations on this planet. The first British units to be seen in the Falkland Islands area in 1982, after the Argentine invasion, were Royal Air Force „Nimrod" maritime reconnaissance aircraft - weeks before the Royal Navy strike force arrived.

By contrast, a modern frigate, such as those deployed today in the navies of NATO partners, is for the most part quite under-equipped, especially in terms of air defense. Although the missile systems themselves are quite effective, the low ammunition supply is a major weakness.

For example, on an Bremen-class frigate (F-122), only eight Sea Sparrow missiles were available for air defense in a launcher on the forecastle, which had to be manually reloaded after the last shot. However, these elder ships are in the process of being decommissioned. A fighter equipped with sea target missiles does not even have to go within range of such defensive weapons. It can comfortably launch its missiles more than 200 km away. A frigate cannot hide. It has only high mobility and the ability to camouflage itself as a saving solution.

The Swedish Visby class is capable of both. The ships can travel at considerable speed. Officially, 35 knots is stated, but one can confidently believe that the Visby is capable of much more. The ship's STEALTH characteristics are provided by its shape, construction material and special paint.

The U.S. Navy recognized as early as World War II that smaller battleships sailing alone could hardly defend themselves against air strikes. Therefore, the FK cruisers were created with the so-called „Aegis system". The ships of the „Ticonderoga" and the „Arleigh Burke" classes are extremely expensive, but also worth every penny.

They can engage aircraft and even now ballistic missiles at a range of more than 150 km and are capable of strategic strikes against distant land targets using „Tomahawk" cruise missiles. Their size allows them to carry more than one hundred missiles. The firepower of two of these ships would probably have been sufficient to defend Pearl Harbor so effectively in December 1942 that not only would the Japanese carrier planes have been shot down long before the islands were sighted, but the Japanese aircraft carrier group would also have been sunk - all at the push of a button. Only fleets of very large states can afford this firepower. So far, only the former Soviet Union, the U.S. and Japan have taken on this expense. The eastern result is also impressive.

The Russian Navy's large cruisers are armed with very powerful missile systems. Nevertheless, Ukraine was able to sink one such guided missile cruiser by firing at it with sea target missiles from land positions. To this day, the U.S. tends to rely on the combat power of its aircraft carriers, which can keep any other fleet at bay. But the effort is enormous.

Without the two smaller aircraft carriers „Invicible" and „Hermes," it would not have been possible for the Royal Navy to prepare and cover the landing of ground troops in the Falkland Islands in 1982. Both carriers brought only a few dozen „Harrier" fighter jets into the fray, only some of which were equipped with radar.

Nevertheless, these jets, which were not capable of supersonic speed, did extremely well. They alone managed the British task force from the Argentine fleet, which had the advantage Argentine fleet, which was able to operate at a short distance from the homeland. Nevertheless, the very short preparation time and inadequate equipment of this campaign ultimately caused the loss of six valuable ships and the death of many men as a result of air attacks.

Combat against land-based forces is particularly dangerous for fleet units because the enemy can use the terrain to advantage to camouflage and secure his forces from detection. In Sweden, a number of bunkers have been constructed in the rocky coastal areas, some of which were designed to provide protection for destroyer-sized ships. Small combat ships can be well hidden from even infrared seekers by camouflage nets and tarpaulins. By rapidly changing position during the night, the enemy must then constantly search in new locations.

Anti-ship guided missiles stationed ashore pose a major threat. They have replaced the classic artillery batteries in coastal protection. Several missile launchers can already be conveniently accommodated on a truck. These missile batteries receive their fire control data from reconnaissance aircraft, drones or radar systems deployed some distance away.

Land forces can still fire against ships with much more diverse weapons. From World War II, it is reported that in at least one case a German „Tiger" tank attacked a destroyer. In the Vietnam War, everything from AK47 rifles to mortars and rocket launchers to artillery pieces was used to attack U.S. coastal forces.

Today, the RPG-7 bazooka is the most popular weapon among pirates off the coast of Africa, because it can easily penetrate the thin outer skin of a ship and is cheap to buy. Even the well known AK47 rifle can easily puncture the steel plates of a hull.

Obviously, a battleship that must operate along enemy coastlines within range of enemy weapons must be equipped differently than a submarine hunter. After the end of the Cold War, submarine hunting became the responsibility of the U.S. Navy's aircraft and submarines. Thus, the development of new high-speed ships for sub-hunting was thrown overboard, and the focus was on a new project called the „Littorial Combat Ship" (LCS).

The new class of ship was to replace the conventional „Oliver Hazard Perry" class frigates. These frigates, designed for air defense and sub-hunting, had lost some of their armament. The ‚standard SM1' missile system had been recognized as too ineffective in the new era - Retrofitting would have been too expensive on an already older ship class of more than 60 units.

The new war on terror actually spurred this development. LCS ships are now not only to support sanctions on foreign coasts, but also to defend their own coasts. To this end, new sophisticated weapons and equipment were devised, for at the same time the LCS was also to set a landmark for technological progress in naval warfare.

For the LCS, stealth against radar was absolutely paramount. Never before had a ship the size of a typical frigate been so rigorously adapted to the anti-RADAR shape so familiar from U.S. Air Force fighters. The two LCS prototypes built are considerably more advanced in this respect than modern developed than modern combat ships planned in Europe, with the exception of the Visby and Skjold classes. Despite some consideration of composite materials, the decision was made not to use the same construction method as for the Visby. The technical limits of plastic construction in terms of strength have not yet been precisely researched and would have been too great a development risk for the very expensive LCS ship type, which was to be more than 100 meters long.

For the design of the LCS, the Navy's contracting command authority had allowed both competition teams to throw almost all traditional technologies and design methods overboard. No wonder, because the ships are to be significantly faster and more assertive than previous designs. Two teams formed in the mid-1990s, made up of various defense and naval industrial firms.

Due to the so-called ‚Jones Act', an old American piece of legislation, no American user is allowed to order new ships outside the USA if they could also be built in the USA. The tender for the LCS showed the established fast ferry shipyards in Australia that they would be entering a familiar technical playing field with this ship, the lightweight construction of large aluminum ships. Austal Ships, currently the world's largest aluminum shipbuilder, took up the challenge and teamed up with the American defense contractor General Dynamics. GD' is one of the most important defense technology manufacturers in the world, which emerged from the ‚Electric Boat Company', the US standard shipyard for the construction of submarines since before the I. World War I.

GD was also an aircraft manufacturer for a few decades, producing the F-16 fighter aircraft, which has since transitioned to Lockheed-Martin. GD provided the electronics, guided weapons and anti-radar technology expertise needed for the extremely challenging LCS project, which the Austal Group, which had previously only manufactured fairly innocuous goods, could not yet boast. Much of the LCS is still top secret. The type ship of this industrial team was given the name „USS Independence." This name had last been used by a decommissioned „Forrestal"-class large aircraft carrier. A sign of the importance attached to the LCS. USS Independence was launched on April 26, 2008.

The second competition team was led by Lockheed-Martin, arguably the largest aerospace company in the world. Lockheed-Martin was able to draw on extensive experience gained from numerous weapons and electronics systems it had built for the U.S. Navy over the years. In addition, they were very experienced in innovative shipbuilding, as Lockheed had also conducted many research programs.

In contrast to the ‚Independence team', they designed a monohull hull because they expected fewer technical problems and better sea performance in very bad weather. Lockheed-Martin also feared that a too unconventional design would scare off the conservative Navy. The type ship was named USS Freedom. It was launched in September 23, 2007.

The cost of the entire program was skyrocketing. So in 2009, the U.S. Navy decided NOT to go with either design. Instead, Lockheed and GD/Austal will be allowed to bid again in an RFP. The winner will then build two units of its type and the loser one.

Salomon would have been proud of such a fair verdict. In the meantime, 13 units of each class are to be built. The „Independence" variant already has four units in service, and four of the „Freedom" have already been released into the world. The extreme imbalance in the U.S. budget at the time probably forced the decision-makers in the White House, Congress and the Pentagon to act in this way.

After all, simply abandoning the project will risk further job losses and thus the loss of further economic power in the U.S. defense industry. For the U.S., defense equipment manufacturing has always been a habitual form of economic stimulus. Whatever the fate of the LCS, it is currently the most sophisticated and advanced military ship in history. Virtually every system and weapon is novel. Never before has a destroyer-sized ship been operated by a crew of only 40 officers, NCO and enlisted personnel. This is a total departure from the philosophies of earlier types of ships, which always handled personnel in a fairly frictional manner. But shouldn't one really ask whether it is really still necessary to assign a sailor to the commander who has nothing more to do than speak his instructions into a microphone? Surely not.

The crew of the LCS is made up of intensively trained technically skilled specialists. You will certainly look in vain for one of the old „tar jackets," whose craft was based primarily on splicing lines and other traditional seamanship skills, on board an LCS. It seems to have been tailored precisely to the younger generations growing up with „smart phones," „I-pads" and computer games.

But it is a mistake to assume that real progress can be made through high mechanization of armaments alone. Recent history has shown that attacks carried out with simple means can achieve an enormously high impact. The primitive AK-47 rifle is still the most common weapon in terms of the historical impact of its use. It has certainly killed millions of people in the sixty years of its existence. Ultimately, in naval warfare history, surprises always occur that completely upset a carefully planned strategy. In World War I, the appearance of submarines was one such unplanned incident. Another was the appearance of seamanlike, well-handled small sailing ships with long-barreled guns that wiped out the Spanish Armada in the English Channel.

The United States has reached the limit of its capabilities after more than a decade of incessant warfare in Iraq, Afghanistan, and elsewhere in the world. If it is not to forfeit its role as the world's leading power, it will be forced to tread much more slowly at certain points than it has in the past.

The new Far Eastern superpowers such as China and India are experiencing high economic growth. It is only natural that they should also develop the necessary self-confidence to play a more important role in the world. The People's Republic of China and India in particular are currently upgrading considerably in the maritime sector. Due to their close proximity and economic competition, the situation between India and China is not without tension. It is possible that the rivalries will one day develop into tangible conflicts. This can happen without Europe or America playing any role at all. However, the concentration of weapons and arms industries in both countries should be kept under observation by the West.

Because in the globalized and networked world, even a local conflict can have glorifying effects. But in the future, even small nations and even guerrilla forces will develop capabilities that will go beyond shooting assault rifles and bazookas as a result of modern technologies.

The appearance of small drone aircraft of unknown origin on Israel's border with Lebanon is a clear sign of this. If one day, not far off, terrorist groups will have remotely controlled unmanned weapons systems, the threat posed by suicide bombers of today will seem small. The miniaturization of computer technology and the good training at Western universities will enable the technicians of such groups to invent completely new threats. These will have an impact not by the mere destruction of the enemy's resources, as in the past, but by their political appeal

# USS INDEPENDENCE LCS-2

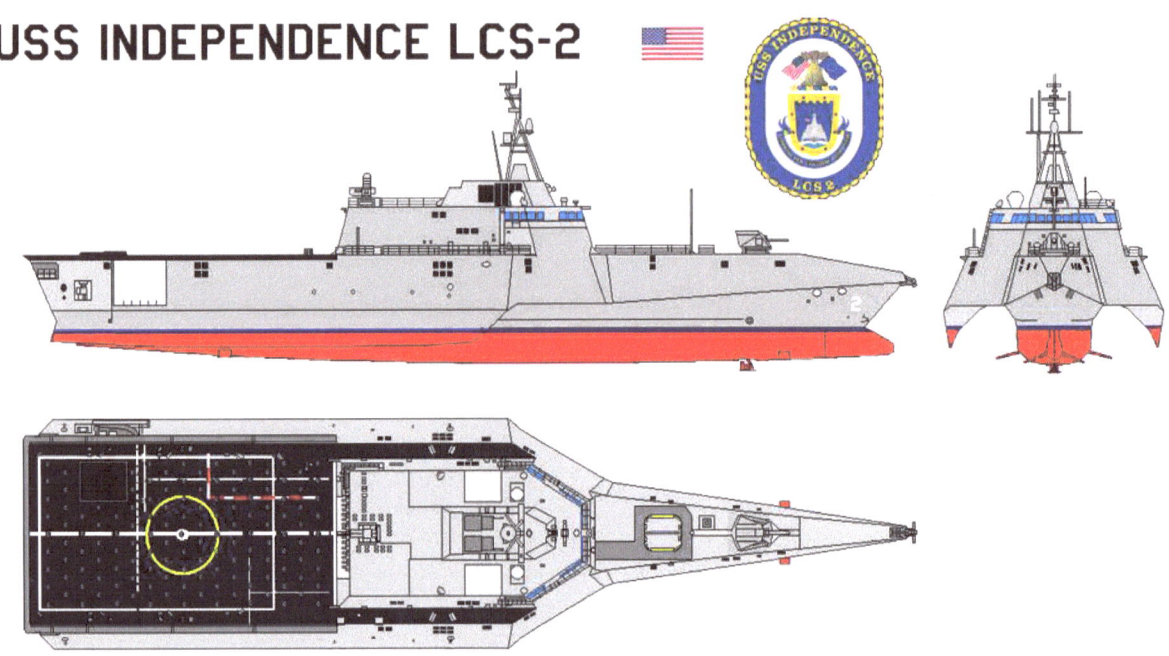

*(Above)* Drawings of the LCS of the USS Independence'. The ship's blocky design owes much to requirements for low radar visibility. As with the Visby-class, all weapons systems and defenses have been sunk into the decks except for the 57mm caliber gun turret, which originated in Sweden. Four large waterjets propel the ship to an officially stated 45 knots, but there is more to come. The hangar is designed for a new type of unmanned helicopter drone to prowl the „hot" zones of a hostile coast. The ship can also carry mine countermeasures, underwater vehicles for combat swimmers and many possible other equipment. At the same time, it can be quickly refitted via containers, depending on the type of mission.

*(Below)* Test detonation of a mine near a LCS to prove the shock resistance of the ship's systems.

*(Above)* The **USS Freedom** - a monohull built by Lockheed-Martin - looks much more conventional than the trimaran. The ship is more conventionally constructed and is made of steel up to the main deck. Nevertheless, it is capable of speeds of supposedly up to 50 knots due to the propulsion system formed by gas turbines and diesels. In any case, this super ship leaves every previous speedboat behind. The electronic equipment is so sophisticated that it is already fully operational with only 40 crew members. It can be used for a wide range of missions, from minehunting to bombardment of land targets to submarine hunting. The guns are still the least important weapons.

*(Left)* The USS Fort Worth at full speed - at least 45 knots.

*(Right)* The start of a soldier's life is always very hard. Here, a LCS is just being thrown into the water.

*(Left) The 'USS Freedom' looks much more conventional than the competition model. This was the intention of the Lockheed Martin-led team, so as not to inundate conservative naval officers with futuristic concepts.*
*After all, such a ship will be driven by a generation that grew up with Playstations and home computers. The new automated systems won't be a problem for these folks, but they might be for the older crew members.*

drones and missiles like the Exocet or the Harpoon. But we can also learn the importance of rapid resupply of land forces. Logistics is the decisive factor in winning a war.

to the public. This was clearly demonstrated in France in 2015, as multiple attacks using low-level resources threw the way of life of an entire nation into chaos. In the future, the missions of naval forces and sea-based police forces will become increasingly intertwined.

NATO's ATALANTA mission to curb piracy off Somalia highlighted not only the technical limitations of today's navies, but also the legal ones. The margins of maneuver to control potential pirates and prosecute them are still too narrow to allow effective action.

But it is not only weapon-bearing combatants that are the focus of navies today. They are also charged today with combating environmental sins, crime, and fisheries protection. So in the future, the naval world will have to rely on new technologies to respond. Perhaps one day, only robotic weapons will fight wars among themselves, or even only virtual combat - ‚gaming' to resolve conflicts between nations

In June 2020 the CNO (Chief Naval. Operations, USN) announced to take the first four LCS out of service. The The reason was high cost of modernization. The concerned vessel are the USS "Freedom", the USS "Independence", the USS "Coronado" and a fourth currently not announce unit. The funds saved will be invested into a new class of more conventional frigates.

It seems that the concept of the LCS have not found so much friends in the US Navy than the creators of the LCS have hoped for. Also critics have been heard about the technical reliability of the complicated vessels technology. On the other hand the concept awaked the interests of Israel, Saudi-Arabia, Malaysia and Taiwan. Any of these nations have rivals quite near to their own shores.

Finally, the LCS does not appear to be as successful a concept for the fleet as the 12 large logistics catamarans built for U.S. forces. The Ukraine war is now teaching us the weakness of surface combatants in the face of remotely controlled air attacks by

*(Above) A Northrop-Grumman MQ-8B „Fire Scout" drone. On the nose of the unmanned helicopter, the sensor container can be seen, which can be swiveled in almost any direction. The cameras it contains allow the remote pilot to see almost anything regardless of the weather or time of day.*
*In addition, a coded laser beam can be aimed at a target, guiding a laser-guided „Hellfire" missile or a laser-guided bomb into a target with centimeter accuracy. The drone can fire „Hellfire" guided weapons itself as well as transmit target data to other units.*

*(Above) Trial run of the LCS-2 „USS Independence". The giant trimaran is capable of speeds exceeding 40 knots, but has much better sea-keeping capabilities than conventional combat ships.*

# The USS „Sea Fighter" - Experiments for the LCS

There was much need in the U.S. Navy for large-scale technical testing before the expensive LCS could be built in series. After all, there was virtually no experience with larger units operating at speeds in excess of 40 knots in littoral areas as part of a superpower's global operations. The U.S. Navy may be the most powerful naval force in world history, but it is strictly conservative at heart. Most of the technologies it employs are much older than 30 years. It doesn't tend to invest a lot of money in reckless experiments without extensive trials. Its budget is also too large for that, because it always involves many billions of U.S. dollars at once, not „just" sums in the triple-digit millions.

So the „Sea Fighter" was developed as a fairly conventional aluminum catamaran powered by diesel and gas turbines. It can achieve long ranges at 20 knots when operated slowly, but can reach more than 50 knots when sprinting. The catamaran has been used primarily to test the new multifunctional containers that will allow the LCS to quickly convert to almost any operation. Also tried was the ability to use such a fast ship as a base for special operations units such as Navy SEALs.

This is because the „Sea Fighter," which is barely visible on radar, can approach a coastline even in shallow waters to which submarines would not have access. To do so, it carries several special SEAL boats on board and can launch and retrieve them via a special ramp system at the stern. No weapons systems are visible on her exterior. But she can unleash armed drone helicopters and, of course, her SEAL soldiers on the enemy - messing with them is a mistake in any case. Thus, the „Sea Fighter" especially well suited for limited warfare, which in the times of the war on terror seems more useful than the use of cruisers or large submarines.

*(Above) The USS „Sea Fighter" in maneuver test at full speed in excess of 35 knots.*

*(Blow) Via an retractable aft ramp a special remote controlled dive vehicle is retrieved on board of the USS "Sea Fighter". Such devices are used for recognizance and mine hunting.*

*(Following page) The anatomy of the „Sea Fighter": (1) jet propulsors, (2) stabilizer fins mounted under the hulls, (3) helicopter landing area, (4) rear vehicle ramp, (5) wheelhouse, (6) nose bead, (7) diesel engines, (8) gas turbines, (9) fuel tanks, (10) boat hangar, (11) crew and Navy SEAL quarters, (12) support systems and storage areas.*

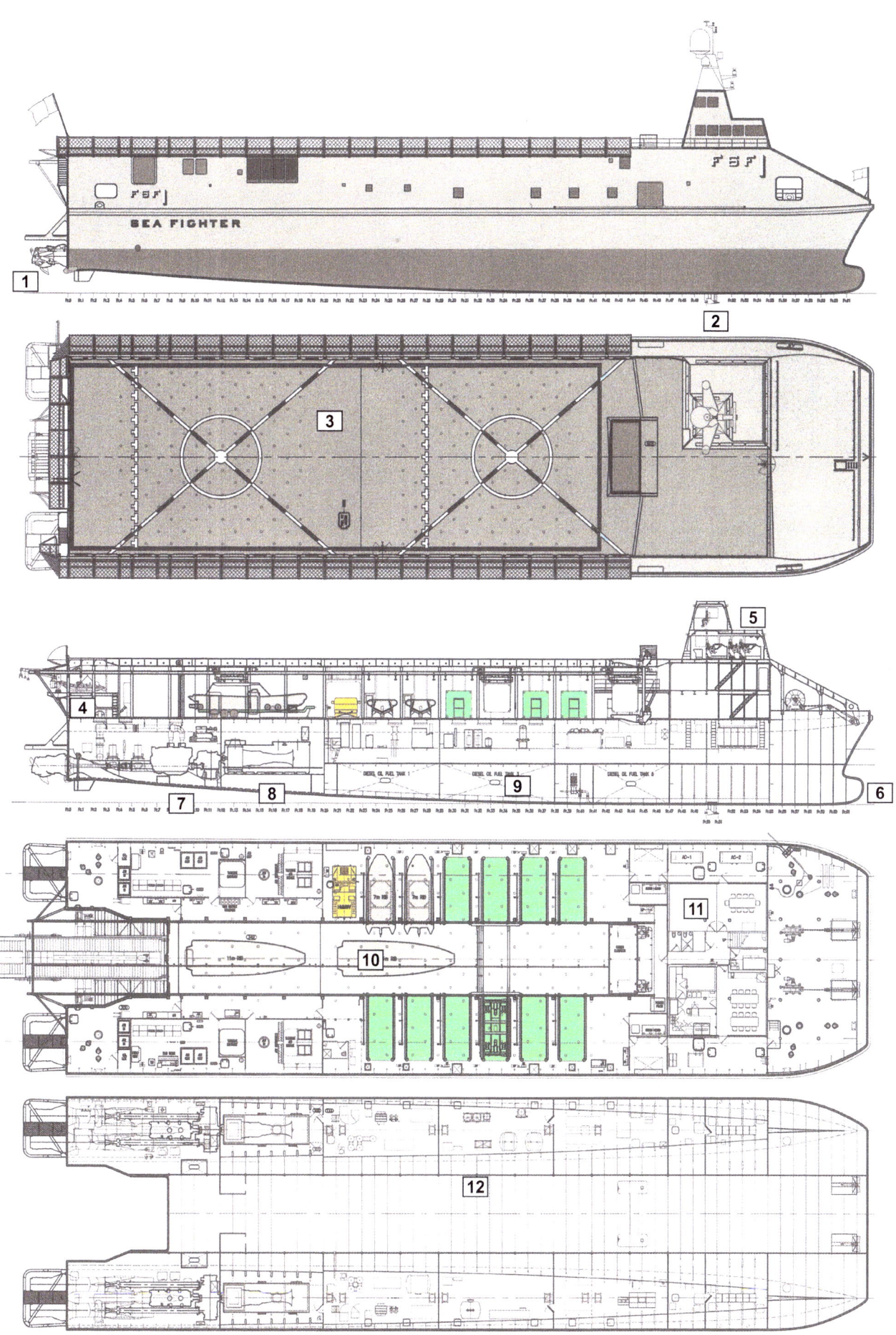

(The „USS Independence" demonstrates its extreme maneuverability at high speed. The waterjet propulsion system is more efficient than propeller propulsion above 25 knots. Thrust redirection provides better maneuverability than previously used blade rudders. The LCS often do not require tug assistance when docking.

Lockheed-Martin's LCS, which looks a bit more conventional,' the USS Freedom on a high-speed run. Apart from the few initiated experts in the U.S. Navy, no one can probably say for sure whether the monohull or the trimaran is the better solution. The decision to build both classes does not make this any easier. But the LCS ships represent the highest level of technology currently achieved in warship construction. The extent of their capabilities is still little known to the public.

# The view into the future

When the first edition of this book was started, the world had just been buffeted in an unprecedented way by the so-called „debt crisis", which caused numerous ups and downs of the securities indices on the world's stock exchanges. It ultimately resulted from the unrestrained indebtedness of many industrialized nations - first and foremost the USA.

Since the political world has now probably realized that even great powers cannot live on „credit" forever, there will be considerable restrictions on government spending worldwide. This will also affect defense spending, which is why the creation of new fleets of large and fast combat ships will be questioned, especially in the United States Navy.

After all, who would be their opponent? It is doubtful whether the Russian armed forces will be able to make up for the emaciation caused by their war of aggression. It is more likely that it will be China, which is growing stronger and more self-confident. History shows, however, that surprising aggressors can always emerge that were not expected. Iraq was such an aggressor in 1992.

Western navies are left with the burden of many small simultaneous tasks that can completely exhaust a navy. Stemming piracy off Somalia's coasts is certainly one of the more important of these tasks. But securing refugee flows from North Africa, pacifying conflicts that break out again and again in West Africa, or providing assistance in the event of natural disasters are also part of this complex.

One can only hope that the modern planners of naval armaments of the future will renounce the gigantism of the past. What is needed are more flexible and cost-effective ship units, such as corvettes, fast transporters and small patrol boats, which are not bristling with weapon systems but can cover as many different tasks as possible. A crane and an dinghy boat may be able to solve a problem more effectively than a cannon.

The threats of our future will not only be aggressors, but also natural disasters, political instabilities, terrorism and ultimately global warming with all its consequences. These challenges will also have to be met on the world's oceans. While the United States, the United Kingdom, and others have some development programs for designing new fast response vehicles, the future of these programs is uncertain or even questionable.

In contrast, China's armed forces are pursuing a very broad armament policy that aims to establish them as a major maritime power in the Pacific as well. The Chinese navy is also present in the Horn of Africa to curb piracy there. The very dense shipyard landscape in China has enabled the navy there to carry out some technically demanding projects, for which the necessary know-how has been taken from civilian fast ferry shipping.

In many places along the Chinese coast, fast catamarans now provide links between cities, villages and industrial areas on the mainland and on islands. The knowledge gained from the construction of the fast ferries benefited the designers of new innovative fast boats and special combat vessels. This was especially true for the processing of aluminum and the use of high-performance diesel engines. A very interesting example is the Type-22 catamaran first put into service in 2004, which is certainly the exact counterpart of the Norwegian Skjold, the Swedish Visby and perhaps also the American LCS.

*(Below) The Houbei class (Type 22) has undoubtedly been influenced by Western STEALTH technologies. Whether it is truly invisible to RADAR remains to be seen. It would take more than a hull shape to make it so. Many design details are derived from Australian fast ferries sold to China.*

This Chinese fast attack craft is an attempt to build up a new threat to Western naval forces, should they wish to interfere in a conflict with the island nation of Taiwan, for example. The disadvantage of such ships is that they are still too slow and too defenseless to stand up to aircraft. Detected by satellites, they will be able to be sunk long before they could get within firing range to attack U.S. Navy aircraft carrier battle groups.

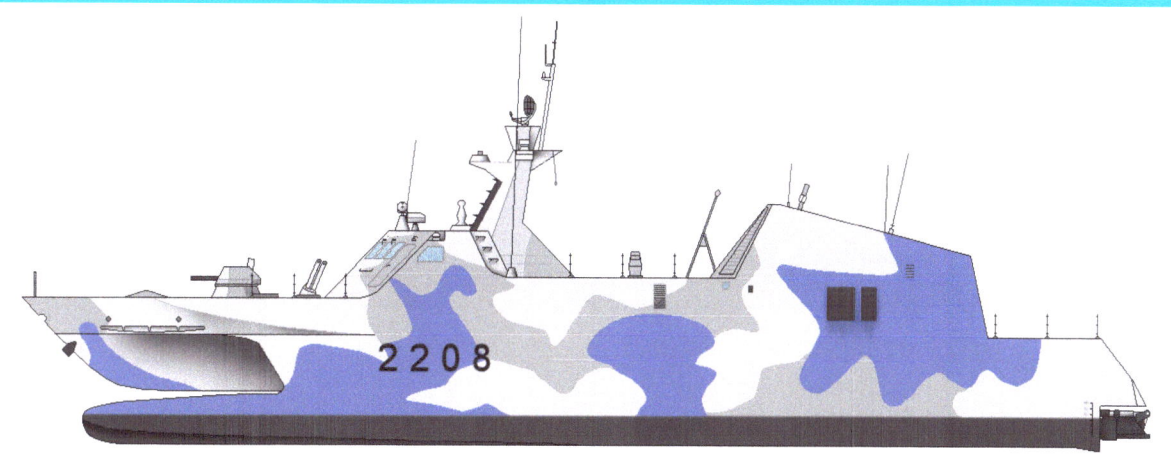

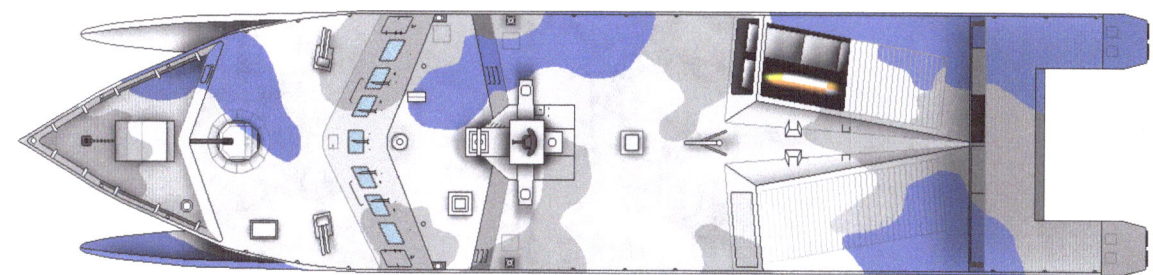

(Left) Paste & Copy is well known in the world of weapon systems. This chinese "LCAC" is a true copy of the American orginal. Many details have been taken to build this prototype sighted on a transport lighter
The class 726 LCAC is a little larger than the American hovercraft ans has been commissioned in 2010. Today about 21 do exist. Surely this craft is part of Chinese plans for an invasion of Taiwan.

# EDAR - New ideas from France

"EDAR" means "Engin de débarquement amphibie rapide" which translates roughly to "Amphibious Fast Landing Craft". THE EDAR is really fast at 30 knots, but it is not a hovercraft. Built as a catamaran, the landing craft is the French answer to the American LCAC, from which the French Navy had bought some to adequately equip its new landing mother ship "Mistral".

The fast catamaran is not much slower than the more expensive LCAC, but can also operate on its own for a few days at only 17 knots cruising speed. This makes it more flexible and versatile in the age of limited operations.

Both tanks, light vehicles or mobile guns can be carried - as well as infantrymen. With a minimum draft of 0.6 meters, it is capable of entering very shallow waters. That is what is special about the EDAR: it is variable in draft. At sea, the platform suspended between the hulls is raised and the draft is increased to 2.5 meters. This makes it very seaworthy and, because of its slender hulls, it can still sail quickly and economically.

The concept of liftable center platforms is not new, but here it helps create a much more economical vessel than the noisy and expensive LCAC could ever be. Also in the UK and even in the USA, cheaper alternatives to the LCAC are being considered - these are very expensive to buy, don't have such a long range and tend to corrode very easily..

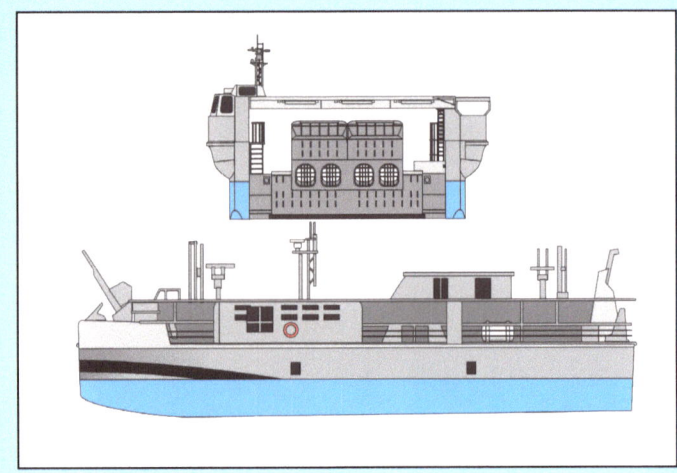

(Above) The front view above shows the middle hull of the EDAR lowered to beach. The blue lower parts pf the side hulls are the wetted area when the craft is running in deeper water.
The EDAR is in both modes seaworthy and stable. This concept is more cost efficient than the hovering landing crafts like the LCAC. They have off course a higher speed, but this would not pretend to be threatened in the same way as slower landing crafts. A landing operation reaquires air supriority and a perfect preparation by bomb and artillery attacks in any case.

(Left and Top) You can clearly see the differences: beach ready to unload and has lowered the center in cruise mode. Its platform is at a safe distance EDAR is compatible with the Mistral-class landing already been exported, it was prevented by the EU also tiny one. Instead, this ship went to Egypt.

*Entering service in 2012, the new Spearhead class is a series of 12 new large catamarans in total, seven of which are already in service. The USS Spearhead has already been deployed off the West African coast to supply humanitarian operations.*

# The far future of military ships

In military, there is a clear distinction between the projects to be realized in the future and the visions presented today by think tanks and designers. They reflect ideas of the conceivable but not yet even the feasible technologies.

Exotic types of weapons, some of which do not exist today, appear, such as railguns, laser weapons or hypersonic missiles. Automation plays a very large role, as the designers mostly belong to generations in which smartphones have taken on a core role in daily life.

Many ideas have already been preconceived by the authors of future novels, comics or by filmmakers. The film „Alien II" by director James Cameron was certainly influential with its ideas about intelligent weapon systems and communicative armor for infantry soldiers. Series like „Star Trek" or games like „HALO" were just as influential. It is simply so that the inventors of the films and the authors make themselves exact thoughts about it, how one can create futuristic as comprehensibly as possible.

But also the will of especially western armed forces to keep human losses as low as possible in the future has led to the introduction of new computer controlled systems like the „Reaper" drones. A remote-controlled combat aircraft is replaceable, but a human being is not. Now, remotely controlled or autonomously operating systems are currently being conceived for surface and undersea warfare. Anti-RADAR technologies are playing a critical role in shaping them. Likewise, drones are being incorporated that can operate under, on, and above the water. The U.S. Navy's LCS is certainly more advanced in this regard than other projects.

High speed will play an increasingly important role. The movement of a fast boat on the ocean is creeping slow compared to that of a fighter aircraft. Thus, the aircraft is always far ahead of it as a weapons carrier. There are always more aircraft than ships, so a belligerent party can more easily tolerate the loss of a few aircraft than that of a frigate or even an aircraft carrier. Therefore, it is doubtful that surface warships will be useful at all in the more distant future. If at all, it will be as floating autonomous drones. Since combat aircraft would then presumably also be replaced by unmanned aerial vehicles, a war of robots would then take place at sea, controlled from some command bunkers on land. The advantage would be that only machines would be destroyed, and not human lives.

An increasingly important role is played by the transport of ground troops and their equipment to distant theaters. Here, the U.S. Navy has taken another step into the future with the new Spearhead-class high-speed transports. After all, the so-called limited warfare, enforcement of sanctions or even disaster relief, which is politically popular today, is best achieved by forces relying on human intervention.

The masses of material and equipment required for this will hardly be transported by aircraft in the future. For example, it has long been common practice for globally operating armed forces to transport soldiers to their destination with light luggage in chartered aircraft, while their material is transported by sea. An M1 tank weighing 67 tons is equivalent to the complete load capacity of a heavy transport aircraft such as the C-5 Galaxy. However, a catamaran like the USS Spearhead can already carry about 600 tons of cargo, or about 8 tanks. The catamaran costs $26 million new; the aircraft would cost over $250 million new. The catamaran crosses the Atlantic to Europe in about 4 days, the airplane needs 1.5 days including handling. There is no question where the cost advantages lie now. Transporters like the Spearhead are more difficult to detect than conventional freighters because of their high speed. Even in the future, the ocean will still be big enough to hide in if you're just fast enough, the key is to avoid danger zones. With the new Zumwalt class, the U.S. Navy has introduced a new generation of ships that looks like pre-World War I vessels, but sets new standards in terms of technology, size and cost. Everything about her has been designed to reduce RADAR emissions. Likewise, the armament is absolutely novel.

Most likely, this class will be equipped with magnetically driven railguns, which should achieve a range of at least 182 kilometers. That would be the longest-range artillery launch ever from a ship. It could reach as far as near-Earth space. In addition, vertically launching guided missiles and cruise missiles from large magazines. The Zumwalt class, which is very large at around 190 meters, does not belong in this book, however, because it is not actually a high-speed combat ship - it „only" travels 30 knots. Perhaps the most far-reaching vision of a future „battleship" is that of the so-called, Startpoint „Dreadnaught 2050". Like the historic battleship of that name, the future one is expected to bring all sorts of innovations to the water around the year 2050. The concept was created before Brexit, so it's questionable whether anyone will be able to afford it then.

The concept envisages a fairly flat large trimaran that is to have hardly any superstructures. Drones are to be used to constantly monitor the surroundings of the ship, which is to be able to reach up to 50 knots. Thus, the equipment with on-board radar is not envisaged. Railguns will serve as armament, with extremely magnetically accelerated projectiles intended to destroy targets by sheer force. Other weapons include lasers for air defense and, again, diversely deployable surface and underwater drones. Submarines are to be repelled by 300-knot torpedoes. These are all technologies that already exist and are being tested today, at least experimentally.

What is really new is the idea of powering the ship by a propulsion system based on nuclear fusion. This is not yet possible today, but intensive work is already being done on the practical application of this type of energy generation. This vision is not as forward-looking as it might seem at first glance. It is merely a matter of technical possibilities that have been consistently taken further and that already exist today.

Nethertheless - It appears that the world's militaries will have to rethink in the not-so-distant future. Artificial intelligence (AI), in particular, will bring the biggest changes. Naval soldiers are probably a phase-out item.

However, there is the question of whether a warship controlled by an AI should be allowed the freedom to fire a weapon on its own decision. There are strong reservations about this vision in wide circles of society and also in the military. However, it has been shown time and again in history that the tendency to always have to obtain better weapons undermines any moral rules. An example of this is the use of armed drones in the fight against hostile terrorist organizations, a low-risk way to fight underground organizations, but one that has cost many innocent lives. Fully automated warfare is as far removed from person- ally fighting a conflict in the manner of the Middle Ages as anything can get. In the final analysis, perhaps, automatic weapons will be directed against all people at the same time, a fight that people would only lose.

A drawing of the USS Zumwalt, a new type of destroyer, which reaches the size of a heavy cruiser of the US Navy from World War II. This would be hopelessly outgunned by the Zumwalt, however, because even one missile hit or railgun shot from her arsenal would destroy her. The old cruiser was operated by a crew of about 1,790; the Zumwalt requires only 158 officers, NCOs and crew ranks.

(Above) „Dreadnaught 2050" is a 2015 British study of a vision of a future battleship. It is expected to boast extreme digitization of all systems, including holographic imaging systems, autonomous drones, and every imaginable futuristic weapon system based on LASER beams. However, fusion reactors are expected to be required for power. Whether these will be available for ships as early as 2050 is written in the stars.

One of the open questions for the future is whether or not surface ships will have a chance to survive on the battlefields. The Ukraine conflict has taught us that a small and not very expensive missile can take out a large and heavily armed cruiser. The sinking of the Russian „Moskva" - the largest active warship in the Black Sea - has shown that the „Moskva" does not have effective air defenses. But most modern warships have this deficiency.

In the future, if artificial intelligence is built into the missile software, it should no longer be possible to effectively combat these sea skimmers and protect surface units. A combination of drone and missile is currently being developed in many navies. This would be the same threat to fleets as the kamikaze fighters on Okinawa in 1945.

# High-speed boats in museums

Few historic high-speed combat ships in the world have found a fate as a museum ship. The nature of the action-packed war effort and the high technical wear and tear on the ships has often prevented their survival. However, some have managed to escape destruction by war or scrapping. It is deliberately refrained from showing all these ships from the lists of maritime museums, but only some special exhibits are presented, which are exemplary or unique, In Wikipedia it is possible under the keyword „List of Museum Ships" to find a comprehensive list of the worldwide inventory of preserved museum ships. Please refer here to the English version, which is very complete.

## The S-71 „Cheetah" („Gepard")

*The „Gepard" is the only western missile speedboat that can be visited. It can be found in the German Naval Museum in Wilhelmshaven (northern Germany). The 48 meter long boat belongs to the Albatross class of the German Navy, built in the 1960s. It can be seen in the condition it was last seen. That means the aft 76 mm gun turret is missing, because the RAM launching facility for anti-aircraft missiles was located here. The "Gepard" is built of plywood, making it somewhat lighter than its French and other Pendants. Most impressive is how much technology was stuffed into this slim and narrow hull. There was just much less room for the crew.*

## The PHM-5 USS „Aries"

The Aries is - as far as known - the only surviving hull of the legendary PHM hydrofoils of the US Navy. It was keel-laid in 1977, placed in service in 1982, and decommissioned in 1993. In 1996, a private initiative saved it from scrapping and anchored it at Grandriver in the US state of Missouri. Despite limited funds, it was restored and preserved. Only the gun turret, the Harpoon launchers and some equipment are still missing today. Today, the USS „Aries" is a unique monument to a post-World War II engineering feat in the United States. She remains the most advanced hydrofoil to date.

## Die HMCS „Bras d'Or"

The „Bras d'Or" is still the fastest hydrofoil in the world to this day, reaching up to 63 knots or 117 km/h during test runs. Today, she stands in Quispamsis in the Canadian province of New Brunswick (east coast). It too is a landmark in the development of hydrofoils and demonstrated the performance potential of this technology. Nethertheless her design was conservative by using propellers and V-shaped hydrofoils.

# The hovercraft „BH7

*The British Hovercraft Corporation's BH7 was probably the first larger hovercraft designed exclusively for the military. It not only served as an experimental model, but was also used in practice. It was largely based on experience with the SR.RN 4 canal ferries, which had already been operating between Dover and Calais for a number of years. The propulsion system was based on the components of these large „floats", but had only a propeller pylon, a fan and a propulsion turbine.*

*The first example floated up for the first time in November 1969 and was tested in various roles as a minesweeper or as a landing craft for troops until 1983, sometimes being used for real missions in the Channel. Encouraged by the spectacular tests, Iran bought two boats of this type while the Shah was still in power. After the revolution in Iran, the influx of spare parts ebbed and these craft lay up in the Persian Gulf for a long time - years ago they could still be easily seen in satellite photos. But they will be mere wrecks by now.*

*The BH7 at the Hovercraft Museum is on display and is one of the showpieces of the collection. The Hovercraft Museum is located near Portsmouth in Southsea - one of the traditional sites of the Royal Navy.*

These are four very special display items, but there are several more. In the U.S. alone, at least four PT speedboats from the World War II era are on display. There is also plenty to see in Europe, Asia and Australia for fans of such ships. But even for „ship spotters" there are always interesting objects to discover - especially if a good zoom lens is used.

*This high speed attack craft of the GDR navy is placed at the maritime museum on the island Daenholm near Stralsund (Germany). It has been an attempted to build cheap and handy attack crafts for the comunist block in the cold war.*

# Tips for model makers

Especially the new and futuristic new battleships should have an appeal to many modelers. However, plans are difficult or impossible to obtain. However, with often a lot of work, the Japanese and other model kit manufacturers have succeeded in creating usable plastic kits of the LCS or other new exotics such as even the LCAC hovercraft. My recommendation, to anyone who wants to build something like this as a remotely controllable vehicle is this: 1. buy such a kit in as large a scale as possible. 2. assemble it, but do not paint it. Also, do not assemble the small parts such as the guns, antennas or other accessories. 3. have the hull 3D scanned by a professional company, keel up and keel down respectively. 4. the resulting 3D model can now be sliced with the appropriate software, giving you frame cracks, water lines and side sections. Likewise, anything that is not quite right can be corrected in the software before building.

The graphic data obtained can now be transformed into a proper model construction plan and printed, just as you are used to doing with other models.

Attention: Remember that the origin of a model kit is subject to copyright protection. A passing on of the plans, the publication in a magazine or even the resale is strictly forbidden with it and punishable. But if you do everything right, you will soon have a model that will steal the show from all others on the model ship pond! It is not an easy thing to design a model hovercraft because of the difficult physics of hovercrafting.

*These classic Matchbox toys are now wanted collector's items in several sales platforms in the internet.*

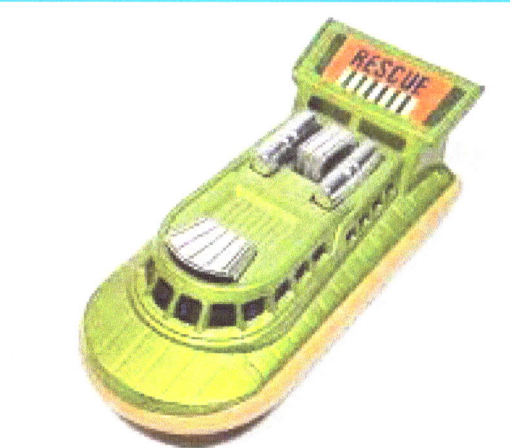

*Standing model of a Pomornik hovercraft in a Russian museum in St. Petersburg. This high quality model has been made just for the museum showcase. A RC-modelhovercraft is not easy to build.*

# Epilogue

I would like to thank the press offices of the German Armed Forces, the Royal Swedish Navy and the United States Navy. In particular, my thanks go to the many photographers and reporters serving with the various armed forces. These servicemen and women, often serving as enlisted or non-commissioned officers, provide the civilian world with an insight into the activities of all those who perform their duties in the armed forces on a daily basis. The reporting reporting often takes place at the risk of personal injury.

I would also like to thank the press departments of the companies that manufacture many of the ships and equipment featured in this book. In particular, I would like to thank the Lürssen shipyard in Vegesack, the Swedish company Dockstarvaret AB, and one of the many press departments within the giant Lockheed-Martin corporation in the United States for permission to use images.

All images used in this book are freely available because they have been kindly provided by the authors or have been declared in the public domain anyway. likewise, the right to use historical free resources applies to some images due to their great age.

*Fast attack crafts equipped with surface-to-surface missiles, automatic multipurpose guns and powerful diesel engines have been for a long period been a constant part of many naval forces. Just the intercontinental operating navies like the US-Navy or the Royal Navy have not made use of those type of crafts.*
*But now the end of their story has come. The German federal navy is currently selling the rest of ther FAC fleet to scrap.*

# Imprint

© 2019 / 2023 by Christof Schramm - All rights reserved including use, storage and publication also in electronic media.

Christof Schramm, 28357 Bremen, Germany

Any mention of names, dates and facts and picture information is done from a consideration of the historical context. This is done on the basis of the sources and data available at the time of creation. Persons shown in pictures purely by chance are members of companies, organizations and armed forces that have released and made available these representations as part of their public relations work.

This edition has been designed to meet the exact requirements of Kindle book publishing (Print on Demand) and therefore cannot be laid out in the same way as typical coffee table books. On the other hand, this edition can be printed in an environmentally friendly way only on demand, which means that there is no need to build up excess inventory.

www.ingramcontent.com/pod-product-compliance
Lightning Source LLC
Chambersburg PA
CBHW051155220526
45473CB00003B/783